Small-scale Water Supply

Small-scale Water Supply

A Review of Technologies

Brian Skinner

PRACTICAL ACTION
Publishing

Practical Action Publishing Ltd
Schumacher Centre for Technology and Development
Bourton on Dunsmore, Rugby,
Warwickshire, CV23 9QZ, UK
www.practicalactionpublishing.org

© WEDC/LSHTM 2003

First published in 2003
Reprinted 2008, 2009

ISBN 978 1 85339 540 6

A catalogue record for this book is available from the British Library.

Since 1974, Practical Action Publishing (formerly Intermediate Technology
Publications and ITDG Publishing) has published and disseminated books and
information in support of international development work throughout the world.
Practical Action Publishing is a trading name of Practical Action Publishing Ltd
(Company Reg. No. 1159018), the wholly owned publishing company of
Practical Action. Practical Action Publishing trades only in support of
its parent charity objectives and any profits are covenanted back to
Practical Action (Charity Reg. No. 247257, Group VAT
Registration No. 880 9924 76).

Illustrations by Ken Chatterton and Rod Shaw
Designed and typeset by J&L Composition, Filey, North Yorkshire
Printed and bound in Great Britain by CPI Antony Rowe,
Chippenham and Eastbourne

Contents

Figures

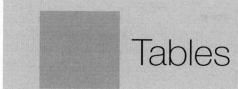

Tables

Boxes

Preface

This book is a replacement for Ross Bulletin 10, *Small Water Supplies*, a booklet which was published in 1978 and was last revised in 1986. Many people have found the Bulletin to be a useful starting point for information about appropriate water supply for rural areas of developing countries, and there is still a demand for it long after it went out of print. The UK government through its Department for International Development (DFID) has financed production of this new version using staff at WELL, one of its Resource Centres.

Since Ross Bulletin 10 was first published, and even since it was last revised, there has been much progress in rural water supply in developing countries. Considerably more published material is now available, giving specific guidance on planning, designing, implementing, operating and maintaining appropriate small water supply schemes using a variety of different technologies.

Like the original book, the purpose of this replacement is to give non-specialists a brief overview of the subject of supplying water to small, low-income communities in rural areas of developing countries. The subject is vast and cannot be comprehensively covered in this short book. Instead, it is expected that the readers who need to design water supply schemes will refer to other published material, or will find assistance from experienced field practitioners.

One of the great lessons of the past twenty years has been the need to pay due attention to the social aspects of water and sanitation, ranging from hygiene to finance. This book cannot do these non-technical topics justice, but the appendices and references mention a number of sources that address these issues and discuss the impact they have on improvements in health and long-term sustainability of water supply schemes.

The book covers most of the technologies found in the original Bulletin but the information has been updated and rearranged. Most of the illustrations are new. Additions to the text include new information about handpumps, water storage, roughing filters, arsenic in groundwater, polyethylene pipes, and improving domestic water using the 'three pot system' and solar disinfection. To restrict the length and scope of this book, some information that was included in the Ross Bulletin 10 has not been repeated in this book. However, comprehensive appendices have been provided to lead readers to other useful sources of information.

At the start of the References, details are given of six 'core books' that cover the whole subject of small water supplies in more detail. Appendix 1 shows which parts of these core books are most relevant to each section of this book. Other specialist books are also listed.

For those who have access to it, the internet is a useful new source of information, one that was never even dreamed of in 1986. Appendix 3 lists sources of information on the worldwide web. Among other things, the WELL website has a searchable library catalogue which can be used to locate the titles of books that cover specific topics relating to water supply.

All measurements in this book use the metric system. Methods for converting these measurements to imperial units are given in Appendix 4.

Details of the original work this book is designed to replace are:
Cairncross, S. and Feachem, R. (1986) *Small Water Supplies*, Bulletin No. 10, The Ross Institute of Tropical Hygiene, London School of Hygiene and Tropical Medicine, London, UK. 78pp. ISBN 0 900995 10 6.

Acknowledgements

Except where shown otherwise, all drawings in this book are produced by the Water, Engineering and Development Centre (WEDC), Loughborough University, UK. Where they are based on drawings from other sources this is shown below. Full details can be found in the references and bibliography.

Figure	Original source
9, 10, 13	Technical Brief 34 in Shaw (1999)
11	Guoth-Gumberger (1987)
15, 37, 38	Morgan (1990)
16, 19	Technical Brief 43 in Shaw (1999)
17	Unknown
23, 24	DHV (1985)
26, 27	Rajagopalan and Shiffman (1974)
28	Nilsson (1988)
29	SWS (1992)
30, 32	Technical Brief 22 in Pickford (1991)
36	IRC (1988)
40, 41, 42, 43, 44, 45	Technical Brief 41 in Pickford (1991)
48, 49	Fraenkel (1997)
52	Nissen-Petersen and Lee (1990)
53	Hasse (1989)
54	Technical Brief 56 in Shaw (1999)
55	Technical Brief 58 in Shaw (1999)
57	Wegelin (1996)
58	IRC (1987)
60	DLVW (1983)
61, 62	IRC (1979)

1 Introduction

1.1 Scope of this book

This book gives an introduction to the technologies that can be used for water supplies in developing countries. It mainly relates to 'point supplies' such as wells, boreholes, springs and rainwater catchment systems that do not use motorized pumps, treatment works, or piped distribution systems. Despite this focus on simple point sources, the book also introduces the reader to powered pumps, water treatment and piped distribution systems.

Most of the point supplies considered here will be suitable only for meeting the water needs of fewer than 250 people. This is because of the low yield of the source or the low delivery rate of the non-motorized method of drawing the water. Some of the technologies, such as rainwater catchment, are best suited to individual families, while solutions such as a gravity-fed piped distribution system from a high-yielding spring may be able to serve a very large community.

This book is too short to provide detailed advice to readers who need to design water supply schemes. They will have to refer to other published material such as that listed in the appendices and references, or obtain assistance from experienced field practitioners. Appendix 1 shows recommended reference material for each sub-section of the book.

1.2 The incremental approach

Improvement of existing supplies is often the best place to start a water project. New schemes may need to be planned incrementally to suit the ability and willingness of a community to be involved and to pay for ever-increasing levels of convenience. Some examples of this incremental approach are suggested in the text.

1.3 The importance of fully involving the community

Experience in the last 20 years has shown that it is not easy to find sustainable solutions to potable water supply for people with low incomes who live in rural areas of developing countries. There has now been a move away from the previous donor, or government, 'top-down' approach of providing a water supply, with very little involvement of the end users, to a 'bottom-up' approach which fully involves them.

The top-down approach often did not adequately meet the real desires of the community. Governments and donors often made a number of false assumptions about what each community wanted, and the way in which the scheme would be operated and maintained. One assumption was that they (the government) could afford to keep the scheme in good running order for the community, but this was not usually feasible, because of shortage of funds to support every community and wrong choices of technology. Where it was assumed that the community would care for the scheme, often in reality it was unable to do so. This was usually because of lack of funds, skill, or the spare parts necessary to operate and maintain the scheme.

As a result of the experience of many failed schemes it is increasingly being realized that, right from the start, there is a need to fully involve the community in the identification of problems, finding possible solutions and agreeing the roles of the community and external agencies in constructing, operating and maintaining water supply schemes. This 'bottom-up' or 'community management' approach is very important to the sustainability of all rural water supply schemes and readers are strongly advised to find out more about the subject from the specialist literature (e.g. IRC, 1991) or field practitioners.

Three important objectives of a water supply scheme are that it should be:

- **acceptable** to the community (e.g. in relation to convenience, and to traditional beliefs and practices) and also acceptable from the environmental and health perspectives
- **feasible** (i.e. suiting the relevant local social, financial, technological and institutional factors)
- **sustainable** (i.e. possible to reliably operate and maintain in the future with the available financial, human, institutional and material resources).

1.4 The importance of sanitation and hygiene

Where there is little understanding of health risks from continued use of existing polluted water sources there may be little willingness to participate in planning, construction, operation or maintenance of a new scheme. If this is the case, then an extensive health and hygiene education programme may be necessary before attempting to start to plan a new/improved water supply. Experience has shown that for the best results such education programmes should be based on participative learning activities and not on listening to lectures. The reader is strongly advised to find out more about this important subject (e.g. Boot, 1991).

The health of a community is not based solely on the quality of the drinking water it consumes. Almost all of the diseases carried by water can be transmitted by other routes. It is therefore important that interventions to improve the health of a community also address the need for improved sanitation and hygiene practices as well as water quality. Too much emphasis is often put on providing pure water at a source. In practice this pure water may become grossly contaminated by poor collection, transportation and handling practices before it is consumed. Even if pure water is consumed, if people have poor sanitation or hygiene practices they may still ingest many pathogens (disease-causing organisms) via dirty hands or contaminated food. Useful guidance on understanding communities' hygiene practices is given in Almedom et al. (1997).

Where water is in short supply, the use of an increased quantity of water will often lead to a decrease in diseases even if the water supplied is not very pure. Convenient access to sufficient amounts of water to allow good hygiene to be practised is therefore also an important factor to consider when designing a new scheme. To ensure appropriate use by a community, a new/improved source must be more convenient than the traditional source.

When planning to considerably increase the amounts of water used by a community it is important to also consider how to dispose of it safely. New health risks (e.g. the breeding of mosquitoes that can carry malaria) can be introduced to the community if used water is not disposed of properly.

2 Design capacity

When designing a water supply, it is necessary to know how much water it should provide. This will vary between countries and it is helpful to measure how much is used from existing improved supplies elsewhere in the region. Table 1 shows some typical ranges of demand that can be used as a starting point if other data are not available.

The amount of water collected from a public water source (e.g. handpump, standpost/tapstand or open well) will be partly dependent on how convenient it is for users to collect the water, although the average demand is often fairly constant for round trip times of between five and 30 minutes.

Convenience is largely dependent on:

- the **round trip walking distance** from the home to the water source
- the presence of **queues** at the source
- the **rate at which water can be collected** from the source
- the **physical effort** (if any) required to lift/pump the water, and
- **whether the user has to pay** for each container of water s/he collects.

The return journey time to the source (including collection time) needs to be less than about five minutes before users are likely to collect sufficient water for a good standard of hygiene. So if their present source is far away a

Table 1 Typical domestic water consumption

Type of water supply	Average consumption[1] (l/person/d)	Range[1] (l/person/d)
Communal water point (e.g. well, handpump or standpost)		
round trip walking distance 500–1000 m	20	10–25
round trip walking distance 250–500 m	20	15–25
round trip walking distance <250 m	25	15–50
Yard tap (i.e. water point outside house but in house compound)	40	20–80
Water point inside house		
single tap	50	30–80
multiple taps	120	70–250

[1] Allowing for 20% wastage
Note: All measurements in this book use the metric system. Methods for converting these measurements to Imperial units are given in Appendix 4.

family will often need a tap located in their yard or house (see Table 1) before their level of hygiene significantly improves. However, a family's level of hygiene is heavily dependent on their knowledge, attitudes and practices and not only on the amount of water available.

A water supply scheme will have a certain design life beyond which it may not be possible to continue without major new rehabilitation works. The likely increase of population during the design life needs to be considered at the design stage or in the future there may be so many people using the source that it becomes congested and long queues develop. If because of financial and other constraints only low levels of supply can be provided initially, then the future plans should strive to allow future staged upgrading. This may make it possible to increase convenience and increase supply per capita when these improvements can be afforded.

Allowance for water used for livestock watering needs to be included in the chosen design capacity if these animals are supplied by the improved source (Table 2). However it is usually more cost-effective if they can continue to use other unimproved sources. Irrigation also needs to be considered. Such use can lead to very high and unsustainable demands, and a community may need to restrict, or ban, the use of water from new sources for irrigation. Used household water can usually be recycled for garden watering. Water demands from institutions such as schools, or for public events such as market days may also need to be considered.

In addition to the estimated demands mentioned above, a factor of safety needs to be included to allow for inaccuracies in estimation and emergencies such as fire fighting. The factor of safety is less important where there are plans for staged improvements since these improvements can be brought forward, or delayed, to suit the actual changes in demand. Also an allowance for wastage, and with piped systems also for leakage, is usually necessary. (Note that the figures in Tables 1 and 2 already include a 20% allowance for wastage.)

If all the demands for the highest demanding day of the week (e.g. market day), and appropriate allowances for wastage, leakage and a factor of safety are added together, this will give the expected maximum total daily demand. This could vary seasonally because of associated changes in water demands

Table 2 Typical livestock water consumption

Type of animal	Average consumption (l/animal/d) allowing for 20% wastage
Cattle	25–35
Horses and mules	20–25
Sheep	15–25
Pigs	10–15
Chickens	0.015–0.025

(From Hofkes (ed.), 1981)

5

or because some alternative sources dry up. Dividing the highest likely daily demand by 24 hours will give the average hourly demand, which needs to be available from whatever source is being used. If the source can only supply water at a rate similar to the average hourly demand then storage will be required to meet peak demands. This is because in practice nearly all of the water will be collected during the hours of daylight and not throughout the day and night. In fact, usually most of it will be required during a few peak hours in the morning, around midday and in the evening. The rate of flow during some of these peak demand periods can easily be from four to six times the average hourly rate of flow, so adequately meeting such demands needs careful consideration. Ways of reducing congestion of people at water collection points during these times also need to be considered (e.g. additional taps on standposts, or additional water collecting devices on hand-dug wells).

Box 1 shows a simple design example to illustrate some of the necessary calculations. Box 9 in Chapter 3 shows the benefits of storage with low-flowing sources.

BOX 1   Estimating demand

Example 1: Gravity supply to taps

Assume that at the end of the design life of a small gravity water supply scheme the population of a village is estimated to be 300. Half of the village are expected to have house tap or yard connections supplying 80 l/person/d and the other half are expected to collect their water from standposts at an average rate of 20 l/person/d. It is not expected that water will be used for livestock or for irrigation. The pipes are polyethylene pipes which are expected to have very little leakage even after many years use, and because of good education and motivation of users in use and repair of the taps the amount of wastage is expected to be only 15%.

The daily demand including 15% for wastage will be:

$$1.15 \times [(150 \times 80) + (150 \times 20)] = 17\,250 \text{ l/day}$$

This is an average daily flow rate of $\dfrac{17\,250}{24 \times 60 \times 60} = 0.2 \text{ l/s} = 11.98 \text{ l/min}$.

This means that allowing for a factor of safety of say 25% a spring supply of at least $(1.25 \times 11.98) = 15$ l/min is required. However storage would be needed to cope with the peak demands occurring during some periods of the day, which on a piped system could easily be four times as much, say at least 60 l/min.

Example 2: Handpump supply

If a gravity supply scheme were not possible, and the 300 people were going to have to use a well or borehole equipped with a handpump, then, because it is less convenient than obtaining water from taps at or near to their homes, the rate of water usage is likely to be less, say an average of 18 l/person/d. The corresponding total daily demand, allowing for 10% wastage will be 5940 l/day.

The rate at which water is withdrawn from the well will depend on the average rate at which users can collect water. If a borehole were used it would need to be able to yield water at the same rate as the handpump. If a hand-dug well is used, then because storage is provided, water can enter the well at a slower rate than that at which the handpump is extracting it (see Section 3.6.1).

The number of hours a pump needs to be used in a day can be calculated by dividing the total daily demand in litres by the pumping rate in litres per hour. This calculation will indicate whether or not congestion is likely at the water point. If to produce the daily demand people have to pump during most of the hours of daylight then it is likely that the community would not find one pump or one source convenient, and an additional pump and/or well/borehole will be required.

For example, if a handpump capable of lifting 18 l/min (1080 l/hr) were used on the well then, to supply the daily demand of 5940 l, the minimum number of hours it would need to be in use would be 5940/1080 = 5.5 hr.

This period of use should be compared with the demand pattern which is acceptable to users. If they want to collect most of their water during 4 peak hours (e.g. 1.5 hours in the morning, 1 hour at midday and 1.5 hours in the early evening) then another handpump may be required.

Instead of adding a second handpump to the same well, providing another well with a handpump at a different location may be preferable, since this would, for many people, reduce the walking distance to their nearest water source. However, note that this improved convenience may mean that the average demand per person may now need to be increased to more than 18 l/person/d. Using two wells would also reduce the depth to which each well has to be excavated below the water table to provide the necessary volume of storage to cope with peak demand, probably making construction much easier.

3 Sources of water

3.1 Introduction

3.1.1 The main sources

An important step in designing a water supply is to choose a suitable source of water. In this chapter the main types of water source are introduced. The rest of the chapter briefly describes how to obtain clean water from each of them. Reference materials that give more detailed information about many of the topics are listed in the appendices and references.

A water source on its own, or in conjunction with other sources, must be able to supply enough water for the community (see Chapter 2). The purification of unsafe water under rural conditions is usually very difficult to achieve because of the cost, non-availability of chemicals and spares, or the high

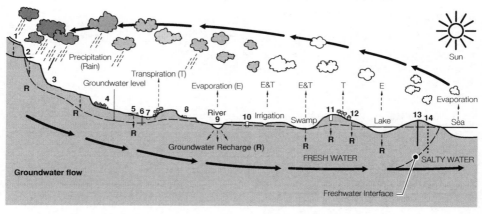

1. Surface water intake	9. Infiltration gallery
2. Spring protection	10. Shallow well
3. Artesian water (not shown)**	11. Pit latrine*
4. Standpost	12. Septic tank*
5. Sand dam reservoir	13. Fresh well
6. Deep well/borehole	14. Salty water well
7. Rainwater harvesting at ground level	* Recharge may pollute groundwater
8. Rainwater catchment from roofs	** Found only in unique situations where groundwater is pressurized

Figure 1 The hydrological cycle, water collection points and groundwater recharge.
Source: WEDC

level of commitment or skills required from a village-level operator. There-fore, **it is usually more appropriate to locate a source that can provide pure water than to attempt to purify polluted water**. Water treatment is discussed further in Chapter 6.

Water is constantly circulating around the earth in what is known as the hydrological cycle (Figure 1). There are three main sources of water: rain-water, surface water and groundwater. These will be examined in more detail in the following sections but Box 2 gives a brief introduction to each of them.

Choosing an appropriate source of water can be very complicated as it depends on many local conditions. Wherever possible it is therefore advisable to seek guidance from someone experienced in this work. Figure 2 is an attempt to illustrate the best approach to the problem, but should not be followed blindly.

BOX 2   Key points about sources of water

When water falls as rain, it runs either in streams or through the ground to rivers which take it to the sea. Water in streams, rivers, ponds and lakes is called surface water, and water flowing underground or emerging at springs is called groundwater.

- **Rainwater:** Rain is usually pure so water for domestic consumption can be collected from the roofs of buildings if they are clean and made of tiles or sheeting, and not of thatch or lead. However, using rainwater collection as the sole source of water is only appropriate for countries with sufficient and reliable rainfall during enough months of the year. If there is a long period without rain, individual storage facilities will need to be large and consequently may not be affordable. Rainwater is a particularly useful source where other sources of water are not available or are polluted.

- **Groundwater** is surface water or rainwater that has soaked into the ground. It is not usually static. Once its downward movement has satu-rated a soil (or fractured rock), the water flows horizontally, often in the same direction as the water on the ground surface. The rate of flow may be very slow. Groundwater is often quite pure because many contami-nants are filtered out as it flows through the ground, particularly if it passes through a fine granular soil. In some types of rock where water flows through cracks and fissures it is not filtered as it flows. In these rocks pollution may carry very far.

- **Surface water** (e.g. from a river or pond) is very likely to be polluted. Its use as a source of drinking water should normally be avoided. If it is to be used, treatment will be required.

It is much better to find a source that provides naturally pure water and then protect it from pollution than to treat water from a polluted source.

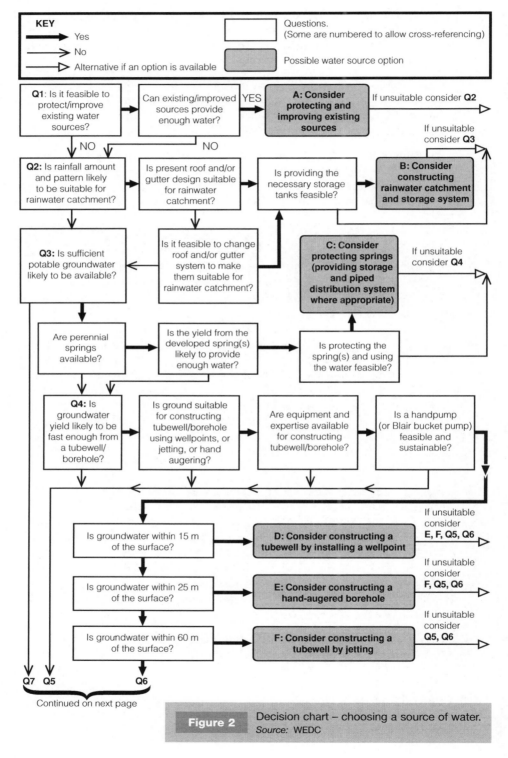

Figure 2 — Decision chart – choosing a source of water.
Source: WEDC

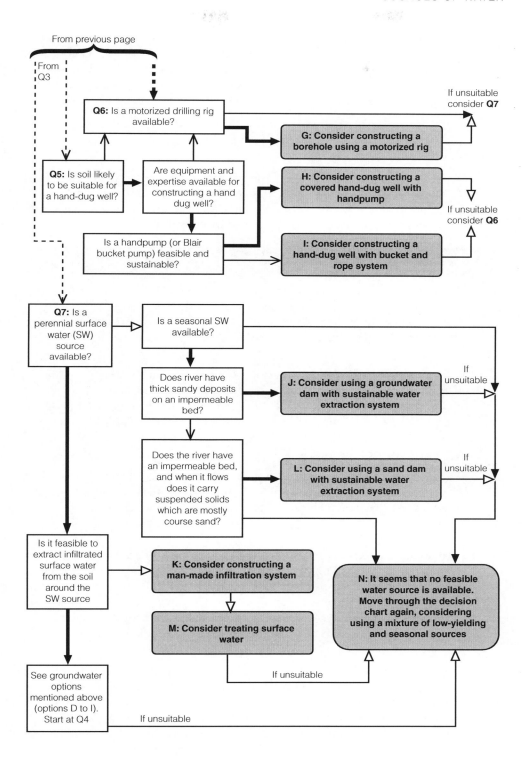

From previous page

From Q3

Q6: Is a motorized drilling rig available?

If unsuitable consider **Q7**

G: Consider constructing a borehole using a motorized rig

Q5: Is soil likely to be suitable for a hand-dug well?

Are equipment and expertise available for constructing a hand dug well?

H: Consider constructing a covered hand-dug well with handpump

If unsuitable consider **Q6**

Is a handpump (or Blair bucket pump) feasible and sustainable?

I: Consider constructing a hand-dug well with bucket and rope system

Q7: Is a perennial surface water (SW) source available?

Is a seasonal SW available?

Does river have thick sandy deposits on an impermeable bed?

J: Consider using a groundwater dam with sustainable water extraction system

If unsuitable

Does the river have an impermeable bed, and when it flows does it carry suspended solids which are mostly course sand?

L: Consider using a sand dam with sustainable water extraction system

If unsuitable

Is it feasible to extract infiltrated surface water from the soil around the SW source

K: Consider constructing a man-made infiltration system

N: It seems that no feasible water source is available. Move through the decision chart again, considering using a mixture of low-yielding and seasonal sources

M: Consider treating surface water

See groundwater options mentioned above (options D to I). Start at Q4

If unsuitable

If unsuitable

11

When a source has been chosen, it should be checked to ensure that it can provide enough water for the community to be served (see Chapter 2). Techniques for measuring the yield of a source are introduced in Section 3.6.2. It may also be appropriate to test the quality of the water, particularly if problems of arsenic or fluoride in groundwater have been reported in the area.

3.1.2 Judging the quality of water

Potential contaminants of water sources are in three categories: pathogens (disease-causing organisms including bacteria, viruses and the eggs of parasitic worms), chemicals (natural and from human activities) and other contaminants. Details of these are summarized in Box 3.

BOX 3   Contaminants in water

- **Pathogens:** These are disease-causing organisms such as bacteria, viruses, protozoa, and eggs or larvae of parasitic worms. Many of these pathogens are carried in human faeces so, wherever sanitation or hygiene is poor, unprotected water sources are at risk.

- **Chemicals:** Agricultural or industrial chemicals cannot easily be removed from water so measures to prevent them ever contaminating a water source are important. Natural chemicals (such as arsenic) which are harmful to humans can be found in groundwater and the presence of non-harmful chemicals (such as iron) can sometimes cause people to reject a water supply. Urine and decomposing faeces can produce chemicals (such as nitrate) that can persist for a long time, particularly in groundwater.

- **Other contaminants:** These include suspended solids and algae (minute water plants). Some of these contaminants (e.g. clay particles) may not be harmful but they may still cause people to reject the water.

Water can be tested to measure the type and amount of contaminants it contains but, as discussed below, this may not be necessary if there are already other indications that it is likely to be contaminated. This is particularly true with bacteriological contamination. Testing for chemical contaminants can be expensive and unnecessary, and professional judgement by someone who knows the region well is advisable when deciding which, if any, tests should be carried out.

Testing water for the presence of any specific pathogen is difficult and expensive so usually tests are carried out to check for the presence of certain types of bacteria which are not necessarily harmful in themselves but which are only found in faeces from warm-blooded animals. Often these bacteria are of the faecal coliform group. These indicator organisms are much easier to measure than the pathogens. Often the *E. coli* bacteria are used as an indicator organism. The concentration of this group of bacteria (i.e. number of *E. coli* per 100 ml of water) is a useful indication of the degree to which the water is contaminated, and the likelihood that pathogens are also present.

Ideally tests should be carried out periodically and especially after treatment processes, to monitor their effectiveness. High quality potable water should have zero *E. coli*/100 ml and this should always be achieved if the water has been properly treated with chlorine. However it is unrealistic to expect even protected, untreated sources in rural areas to always achieve zero *E. coli* /100 ml. Many practitioners consider levels of up to 10 *E. coli* /100 ml to be satisfactory. Sources with contamination levels above 50 *E. coli* /100 ml definitely need urgent investigation to find ways of reducing contamination. If this is not possible then an alternative, purer source of water should be sought or treatment of the contaminated water should be considered (Chapter 6).

There are now some good field kits available for both chemical and bacteriological testing of water. However, field and laboratory tests are likely to be reliable only if a well-trained person carries out the correct sampling and testing procedures. Some of the tests can be quite expensive.

As mentioned in Section 1.4, many water-related diseases can be spread by other routes, and clean water will often be contaminated after collection. Hence, attention to improving water quality at a source also needs to be accompanied by interventions to improve household sanitation and hygiene.

Instead of testing the water, a sanitary survey can be carried out to assess the potential risks of the water being polluted by human activities. If major risks are identified, then it is not really worth carrying out chemical or bacteriological tests until these risks are removed. Good advice on sanitary surveys is found in Lloyd and Helmer (1991). Laboratory testing only gives the situation at the time of the test, whereas a sanitary survey can reveal conditions or practices that may cause intermittent pollution incidents which might otherwise be missed by periodic sampling.

Key issues relating to the quality of water from untreated sources are presented in Box 4 (overleaf).

3.2 Rainwater

An impermeable catchment surface, a collection system and a storage container are needed to capture rainwater for domestic purposes. Raised catchments such as roofs can produce good quality water. Ground-level catchments (such as threshing floors or expanses of rock, etc.) can also be used but the quality of the water is likely to be much poorer. Using rainwater is particularly appropriate in the situations mentioned in Box 5.

Rainwater for potable purposes can be collected from roofs made of tiles, thin flat pieces of rock (e.g. slate), or sheeting (galvanized iron, aluminium or

---

BOX 4 Quality of water from untreated sources

- It is likely that there will be contamination of untreated sources. It is necessary to work with the community to reduce the risk of contamination to a minimum by good design, appropriate water collection methods, and good sanitation and hygiene practices.

- Water quality testing is not always necessary. A visual inspection (e.g. a sanitary survey) can often be used to indicate whether a source is likely to be contaminated by human activities. The survey can also identify probable sources of pollution.

- Domestic use of sufficient quantities of water which is not pure will usually lead to greater improvements in health than if the family has only a small quantity of high quality water.

- Users need to understand how to avoid contaminating water during and after collection. There is little benefit in providing good quality water at a source if it will subsequently become polluted before it is consumed.

BOX 5 Key points about using rainwater

Rainwater catchment is particularly appropriate where:

- there is a suitable pattern of rainfall during the year *and*
- the community wants to use it *and*
- people can afford the system *and* are able to maintain it

and, additionally where other sources of water are:

- not available *or* only seasonally available *or*
- polluted *or*
- inconveniently located *or*
- unreliable.

---

asbestos cement). Water from roofs that are partly covered with lead may contain harmful levels of lead, so should not be consumed. Thatched or mud roofs, or bituminous surfaces (such as roofing felt) are likely to make the water unpleasant to taste. Thatched roofs can also encourage rodents that may deposit pathogens that are subsequently washed off in the rainwater.

Water running off a roof is usually collected in a gutter that runs along the edge of the roof (Figure 3). The gutter should slope evenly towards the downpipe that serves the tank. If the gutter sags, pools of water can be left after rain. These can form breeding places for mosquitoes. One simple

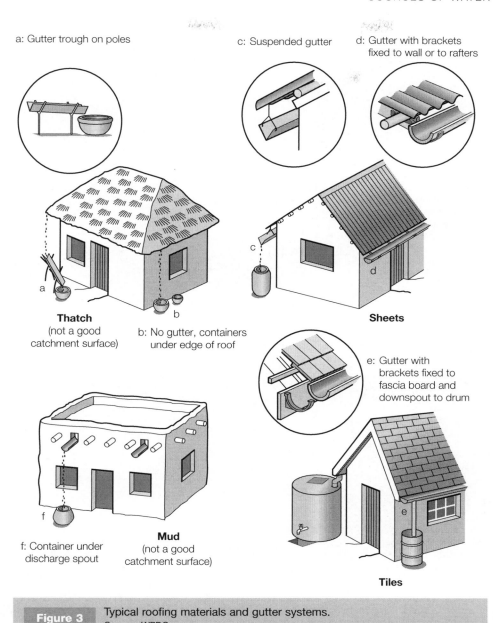

a: Gutter trough on poles

c: Suspended gutter

d: Gutter with brackets fixed to wall or to rafters

Thatch
(not a good catchment surface)

b: No gutter, containers under edge of roof

Sheets

e: Gutter with brackets fixed to fascia board and downspout to drum

f: Container under discharge spout

Mud
(not a good catchment surface)

Tiles

Figure 3	Typical roofing materials and gutter systems. *Source:* WEDC

design of gutter is the suspended 'V' gutter. If used with a deflector plate a suspended gutter can be positioned much lower than a conventional gutter (which is fixed to the fascia board) and yet it can still collect most of the water running off the roof (Figure 4).

Dust, dead leaves and bird droppings may fall on the roof during dry periods, and be washed into the gutter by the first rain. They should therefore be

15

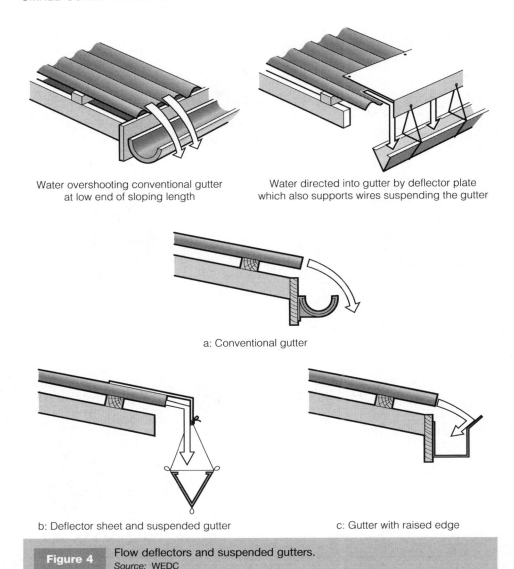

Water overshooting conventional gutter at low end of sloping length

Water directed into gutter by deflector plate which also supports wires suspending the gutter

a: Conventional gutter

b: Deflector sheet and suspended gutter

c: Gutter with raised edge

Figure 4 **Flow deflectors and suspended gutters.**
Source: WEDC

cleaned off just before the start of the rainy season. An inclined wire mesh positioned under the end of the downpipe, just before it enters the tank, is a useful way of preventing debris entering the tank, and such a mesh is largely self-cleaning (Figure 5). Another option is to arrange the pipework so that it can be temporarily detached from the water storage container, to waste the initial flow of water which is likely to carry most of the pollution. This method had the disadvantage that it needs someone to operate it while it is still raining. One method of diverting the first flow that does not need someone to work out in the rain is to use a sump pipe (Figure 6). The sump pipe needs to be emptied after each period of rainfall.

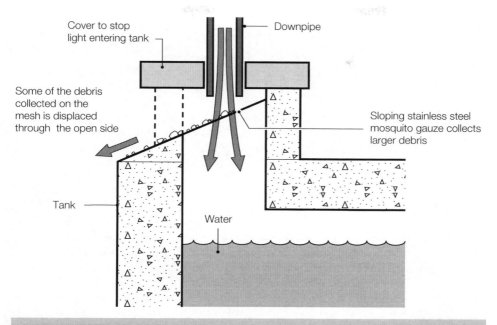

Cover to stop
light entering tank

Downpipe

Some of the debris
collected on the
mesh is displaced
through the open side

Sloping stainless steel
mosquito gauze collects
larger debris

Tank

Water

Figure 5 Angled mesh strainer used to intercept larger debris in captured rainwater.
Source: WEDC

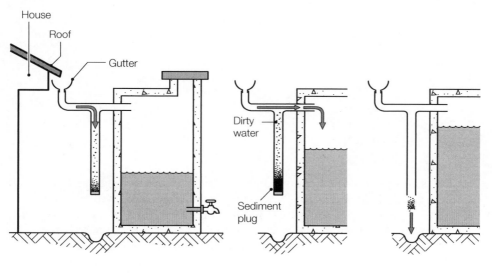

House

Roof

Gutter

Dirty
water

Sediment
plug

a: The first flow of captured rainwater
and any suspended debris enter
the sump rather than the tank

b: Later flows, which
should contain much
less debris, pass into
the tank because the
sump is full

c: Removal of the
plug cleans out the
water and sediment
in the sump
preparing it for the
next storm

Figure 6 A sump-type of 'first flush' diverter.
Source: WEDC

BOX 6 Basic calculations for rainwater catchment

To estimate how much rainwater can be collected from a roof a designer needs to know the area of the roof in plan and the yearly rainfall. One millimetre (mm) of rainfall on one square metre of impermeable roof (e.g. tiles) will give about 0.8 l of water, after allowing for evaporation and other minor losses. So, if the roof measures 5 m × 8 m in plan, and the average annual rainfall in dry years is 750 mm, the amount of rainwater which can be collected in a year is equal to

$$5 \times 8 \times 750 \times 0.8 = 24\,000 \text{ litres per year or } \frac{24\,000}{365} = 66 \text{ l/day on average.}$$

This can be compared with the expected requirements of the people who will use the water to see if rainwater is an appropriate source. If it will be enough then the next stage is to decide on the amount of storage required.

An initial guide for the storage required is to calculate the amount of water which will be needed during the longest period when there is no rain. For example if a household needs a total of 35 l of drinking water per day, and there is a dry season of 4.5 months (137 days) the minimum volume of storage needed at the start of the dry season is:

$$137 \times 35 = 4795 \text{ l or } 4.8 \text{ m}^3$$

In this case, allowing for a factor of safety, it would probably be appropriate to store at least 5.5 m^3.

A more detailed, month by month, analysis of the rainfall captured and used during the period leading up to the dry season should also be carried out. This is to check that when the rains stop, enough water is stored to provide all the water needed before it next rains.

It is convenient to keep rainwater for domestic use in a storage tank beside the house. If the tank is above ground it can be equipped with a tap to facilitate hygienic collection of the water. The affordability of sufficient storage for water for all domestic needs is often a problem. Storage of only the water used for drinking and food preparation is more realistic and practical if sources of lower quality water are available for other domestic uses (see Section 5.3 for various methods of storage). As long as the tank is covered, and the water does not contain much organic material, its quality will improve during storage (see Section 6.3).

3.3 Groundwater

3.3.1 Introduction to groundwater and methods of reaching it

As mentioned in Box 2, groundwater is often quite pure, particularly when it has passed through fine granular soils. Often 2 m depth of fine soil, when unsaturated, will successfully remove virtually all pathogenic organisms. If polluted water enters coarser soils (such as gravel), or passes into fractures in rock, filtration does not take place. However, many pathogens will still die after a few weeks because of the low temperature and an absence of nutrients. Occasionally natural chemicals found in soil and rocks can create high concentrations of fluoride or arsenic in the groundwater which make it unsuitable for human consumption. In other situations, groundwater can contain high concentrations of iron or manganese, which, although not harmful, can result in people rejecting the water because of its taste or colour, or the way it stains food or laundry. Also, some groundwater can be quite acidic causing steel pipes to corrode quickly, even when protected by galvanizing.

To reduce the risks of contamination, the source of groundwater (the well, borehole or spring) should be as far away as possible from concentrations of harmful bacteria and viruses such as pit latrines and septic tanks, especially where these are on the uphill side of the source. A minimum separation distance of 30 m is often suggested. The safe distance will depend on the depth and type of soil above the groundwater, since this will control the effectiveness of the natural filtration process. The speed of the movement of the groundwater is also relevant. Where the time of travel of liquids that might contain viruses and bacteria exceeds 25 days the health risks are low, and beyond 50 days travel time the risks are very low. A separation of less than 30 m is likely to be appropriate where there is a good depth of unsaturated fine soil between the pit and the level of the groundwater. Where the pollution could quickly enter fractured rocks the safe distance would be much more than 30 m.

Ideally the quality of a good source of groundwater should be monitored to check that that it does not deteriorate over time. New development near the source may need to be restricted to protect the quality of the water.

A formation of soil or fractured rock from which water can be extracted is called an aquifer. Other soils, like clay, are not aquifers because although they may contain water it can not be easily extracted because it is held tight to the very fine particles.

The level to which the water saturates an aquifer is called the groundwater table. In a uniform granular soil the groundwater table may be level over a large area and it is not particularly important where a well or borehole is sited. However, in a fractured rock formation the water level may vary from place to place depending on the number and pattern of the fractures. For such aquifers the positioning of a borehole is much more critical to achieve a good yield of water.

If, due to the slope of the ground, the water table in an aquifer reaches ground level the groundwater will come out of the ground as a spring (Figure 7). If the spring has sufficient flow throughout the year it is often an ideal source of water since, unlike water from wells and boreholes, it does not have to be lifted, or pumped, to the surface. Section 3.3.2 gives more information about ways to collect water from springs.

Where a spring is high enough above a community the water from it can often be distributed by gravity through pipes to taps conveniently positioned at a number of points close to the users (Figure 63).

Where there are no springs then the next best alternative is usually to reach the groundwater using a dug well (Section 3.3.4) or a borehole (Section 3.3.3).

Where an aquifer occurs between two impermeable rock or soil formations it is called a confined aquifer (see Figure 8). An artesian aquifer is a fully saturated, confined aquifer in which the groundwater is pressurized so that, if a borehole penetrates it, water will rise above the top of the aquifer. Sometimes the pressure will cause it to flow out of the borehole above ground level. This is known as an artesian borehole. A similar feature is naturally present in an artesian spring, which occurs when a fissure penetrates through the upper confining layer (Figure 8).

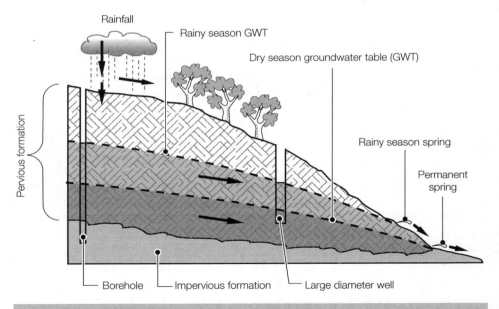

Figure 7	A well, borehole and gravity overflow spring.
	Source: WEDC

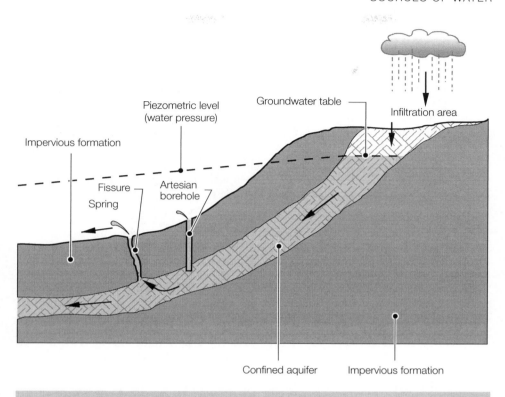

Piezometric level
(water pressure)

Groundwater table

Infiltration area

Impervious formation

Fissure

Spring

Artesian
borehole

Confined aquifer   Impervious formation

Figure 8	Artesian fissure spring and artesian borehole.
	Source: WEDC

BOX 7   Key points about using groundwater

Advantages

- Often it is fairly pure although this may not be the case with groundwater from coarse soils or fractured rocks, or where certain natural chemicals are present.

- Often it is available all year round, close to communities, and in considerable quantities.

Disadvantages

- It can be expensive to access the groundwater, particularly if it is deep below the surface.

- Except in the case of natural springs and artesian boreholes, groundwater requires a lifting device which needs to be acceptable to, and sustainable by the community.

Before it is decided to dig a well or construct a borehole, it is necessary to find out as much as possible about the groundwater in the area. The best place to site a well or borehole is where the water table is nearest to the ground surface. Near to rivers, ponds and springs, and in low-lying areas (Figure 1) groundwater is usually close to the surface. However, where it is at shallow depth there is also the risk that it can become polluted by human activities on the surface. Certain types of tree or, in dry seasons, a richer growth of vegetation can be signs that water is close to ground level. Looking at the level of water in existing open wells and talking to the people who have built them is also helpful. Try to find out about the local soil conditions. A layer of rock, for instance, may make well-digging or hand-augering of boreholes impossible. It is usually easiest to extract water from gravels and sands.

Once it is decided that a groundwater supply is appropriate, there are five main methods which can be employed to reach it. These are:

■ driven tubewell
■ jetted tubewell
■ hand-augered borehole
■ machine-drilled borehole
■ hand-dug well.

Each technique has various advantages and disadvantages. These are briefly discussed below. More detailed examination of each option may be found in the Section 3.3.3.

A **driven tubewell**, made by hammering a special pipe into suitable ground, is one of the easiest types of well to install. Unsuitable ground includes heavy clay or rocky soils, or sands more than 10–15 m deep. The pipe needs a special filter, called a wellpoint, at its lower end and this is unlikely to be available locally. Driven tubewells are not often used.

A **jetted tubewell** is formed when pressurized water is used to excavate a narrow hole into the soil. To construct this type of tubewell, water is pumped down a pipe that is gradually pushed into the soil. This method is suitable mainly for uniform fine-grained soils in which it can be used to sink wells up to 80 m deep. The sludging method of excavation of a tubewell (Figure 19) does not need a pump to pressurize the water. Instead it uses reverse jetting to remove excavated soil.

A **hand-augered borehole** is a simple type of borehole. It can be drilled relatively cheaply in sandy soils to a depth of about 25 m in a few days. A tripod and hand-operated winch are usually required to raise the drilling stem and soil-filled auger during construction.

A **machine-drilled borehole** is usually the most expensive way of accessing groundwater, particularly if large drilling rigs are used. The larger rigs can drill through many metres of rock relatively easily, to reach depths of over 100 m. However if the groundwater is over 45 m deep, it may be difficult to find a suitable handpump to raise it. Even very deep boreholes may not find water so it is best if an expert is involved in choosing where to site the hole.

In recent years small, simple drilling rigs, which use a mixture of human and engine power have become available. These have considerably reduced the cost of boreholes. Some of these smaller rigs are very portable, capable of being carried in the back of a pick-up truck and then being manhandled into place.

A **hand-dug well** can succeed in places where a tubewell or borehole cannot. This may be the case where water is present in the ground but will only seep into a tubewell very slowly. A hand-dug well is of considerably larger diameter than a tube well, so that more water can seep in over the larger surface area. However, its main advantage is that it has a large storage capacity. Often this means that a lot of water can be drawn from it during the day without it running dry, and can be replaced by groundwater that flows into the well at night. However, digging and lining a well by hand can take a great deal of effort – about five man-days per metre of depth. It can also be dangerous for inexperienced workers. It is therefore suggested that wells should only be constructed where one of the other methods is not feasible and there is good reason to expect water within 40 m of the surface.

A major advantage of a hand-dug well is that water can be drawn from it using the simple rope and bucket method (Section 4.2.1). The narrow boreholes previously mentioned usually require a handpump, which may be difficult for a community to sustain. Section 4.2.1 describes a narrow valved bucket (the Blair bucket) which may be a feasible alternative.

Unlike many of the techniques used for boreholes it is not normally possible to construct a hand-dug well very far below the groundwater table. Excavation below groundwater should therefore take place at the end of the driest season of the year, when the water table will be at its lowest. Manual excavation should be unhampered by water so an effective method of groundwater removal from the well, faster than it enters, is required during excavation.

Table 3 shows the main advantages and disadvantages of hand-dug wells compared to hand-augered and machine-drilled boreholes.

3.3.2 Springs

Spring water flows from an aquifer in which the water table reaches ground level. Often this will occur when a layer of impermeable material below the aquifer is at or close to the ground surface (Figure 7). Alternatively springs can occur where water-filled fractures in a rock reach the surface.

The best places to look for springs are on the slopes of hillsides and in river valleys. Green vegetation at a certain point in a dry area may indicate a spring, or one may be found by following a stream up to its source. The local inhabitants, particularly the women, are usually the best guides, as they know most springs in their area and whether the spring flow is steady throughout the year or if it tends to dry up towards the end of the dry season.

Spring water is usually pure, but it can become polluted if it stands in an open pool, or flows over the ground. A spring should therefore be protected in some

Table 3 A comparison of wells and boreholes

Hand-dug wells	Hand-augered boreholes	Machine-drilled borehole
✔ A flexible procedure which can be adapted to a variety of soil conditions as long as lining materials are available.	✗ Not usually suitable where there are large stones or hard layers.	✔ Equipment is available for all types of ground to great depths.
✔ Often only cement, some reinforcement, and locally available materials such as sand and aggregate (or bricks) are needed for the lining.	✗ Needs borehole casing pipes.	✗ Needs borehole casing pipes.
✔ The resulting wide-mouthed well is easily adaptable to simple water lifting devices, some of which allow more than one person to draw water at the same time.	✗ Needs a sustainable handpump, footpump or Blair bucket pump. Only one person at a time can collect water.	✗ Needs a sustainable handpump, footpump or Blair bucket pump. Only one person at a time can collect water.
✔ The well provides a reservoir which is useful for collecting and storing water from ground formations which yield water slowly.	✗ The low storage volume in the borehole means it is unlikely to be suitable for a low-yielding aquifer.	✗ The low storage volume in the borehole means it is unlikely to be suitable for a low-yielding aquifer.
✔ In some situations the large number of members of the community who can be involved is seen as an advantage since it can lead to an increased sense of 'ownership' of the well. The method can use unskilled labour working with a few trained workers.	Possible disadvantage: Requires few people so there is less opportunity for full community involvement and the associated sense of ownership.	Possible disadvantage: Requires only a few people to operate the rig, and generally there is little opportunity for community involvement.
✗ Takes a long time to construct.	✔ It is relatively fast.	✔ It is very fast.
✗ Uses a lot of bulky, heavy materials for the lining that may need to be transported in a lorry.	✔ Far less equipment needs transporting to the site.	✗ Heavy rigs require the construction of special access roads to the borehole site. They may not be able to travel in the rainy season.
✗ It needs a dedicated team of people willing to work hard.	✔ Needs fewer people and the work is not as hard.	✔ Needs very few people. The machine does most of the hard work.
✗ Construction is usually more expensive than a hand-drilled borehole of the same depth.	✔ Often cheaper than well.	✗ Usually much more expensive than well.

Table 3 (Continued)

Hand-dug wells	Hand-augered boreholes	Machine-drilled borehole
✗ Hand-digging cannot easily penetrate hard ground and rock, although this may be possible if air compressors and jack hammers (with explosives if necessary) are used.	✗ Difficult to penetrate hard ground.	✔ Equipment is available for all types of ground to great depths.
✗ In some soils hand-digging below the water table is not easy. This problem can be minimized by digging the last part of the well at the end of the dry season, but even then, because of the entry of water, it may be hard to dig far enough below the water table to guard against the well running dry during years of drought.	✔ Can penetrate relatively deep below the water table even in loose sandy soils, so the borehole can be constructed deep below the water table.	✔ Can penetrate relatively deep below the water table even in loose sandy soils so the borehole can be constructed deep below the water table.
✗ Open-topped wells and the use of simple water lifting devices have associated pollution risks. These can be avoided if the well has a cover slab and handpump but then only one person can draw water at a time unless more than one handpump is fitted. The handpump needs to be sustainable to ensure continued water supply (a locked access hatch can be provided for a bucket and rope to be used in an emergency).	✔ Less risk of pollution of groundwater by users during collection.	✔ Less risk of pollution of groundwater by users during collection.
✗ Difficult to make very deep (often not > 20m, very rarely > 30m).	✗ Difficult to make deep using hand-augering (rarely > 30m except by jetting with drilling mud in fine sands, when depths of over 60m can be achieved). Cannot usually be deepened.	✔ Can be constructed to great depth.
✔ Can usually be deepened if necessary because water table falls unexpectedly.		
✗ Some safety hazards to diggers (and to users if left as an open well).	✔ Negligible safety hazards.	✔ Limited safety hazards.

Key: ✔ = Advantage; ✗ = Disadvantage.

25

way, so that the water can be collected without being open to surface pollution. The three main objectives of protecting a spring are indicated in Box 8.

It is important to check that the spring is formed by water seeping from the ground and not from a stream or rainwater runoff that has merely gone underground for a short distance. Observing the flow and temperature of the spring water shortly after rain can test this. If after a heavy rainstorm the temperature and flow rate of the spring quickly change it indicates that there is a risk that some, or even all, of the spring water may be surface water, which will not have been sufficiently purified by natural filtration through soil.

The yield of a spring can be checked as described in Section 3.6.

To protect a spring from potential contamination, it is advisable to dig back into the hillside along the water-bearing layer, to capture water from below ground. The deeper below the surface, the more protected the water will be from surface pollutants. If flow is concentrated at one point (the eye of the spring) walls of clay, concrete or brickwork can be used to intercept the flow and to channel it into a pipe or stone-filled trench cut into the impermeable layer (Figure 9). The pipe or stone-filled trench is used to carry the water to

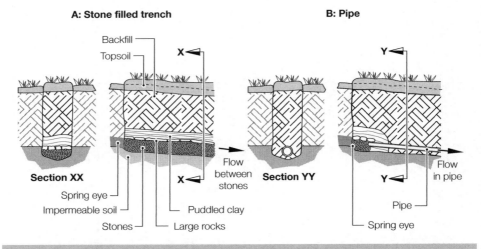

Figure 9   Two ways of collecting and conveying the water from the spring eye.
Source: Shaw (1999)

26

a suitable discharge point such as a headwall with an apron slab and drain (Figure 10) or to a storage tank.

A filter of graded stones is usually necessary at the eye of the spring to stop the surrounding soil being eroded. The pipe or stone-filled trench from the eye must be large enough to carry the highest rate of flow experienced during the year. If this is not desirable a separate overflow pipe, or another stone-filled trench, should be provided near the eye, to prevent the water level in the aquifer ever rising above its natural level.

If there are several different eyes, water from each can be channelled into separate pipes which can all discharge into a small inspection chamber,

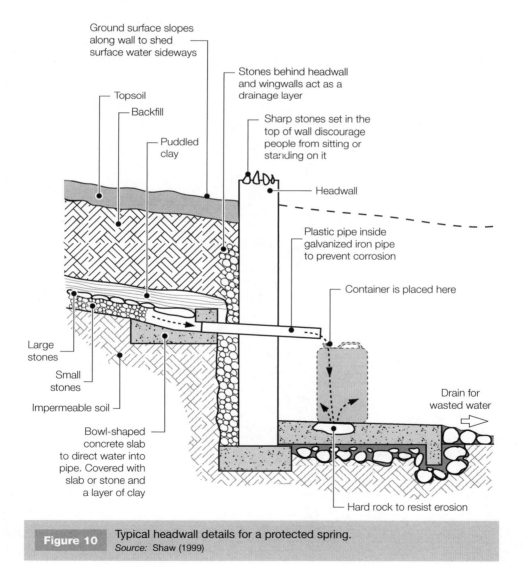

Figure 10	Typical headwall details for a protected spring.
	Source: Shaw (1999)

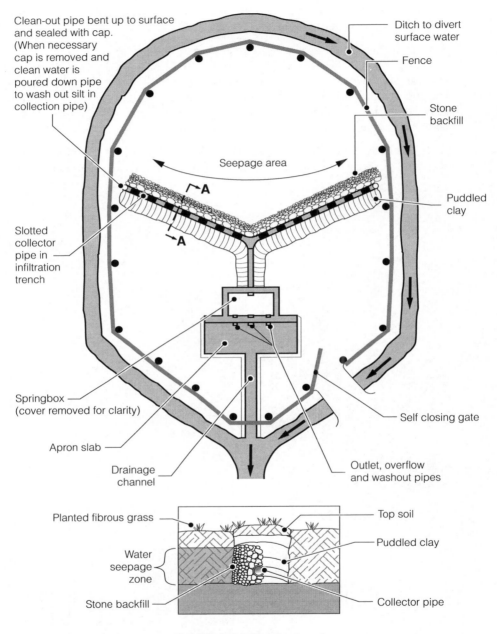

Clean-out pipe bent up to surface and sealed with cap. (When necessary cap is removed and clean water is poured down pipe to wash out silt in collection pipe)

Ditch to divert surface water

Fence

Stone backfill

Seepage area

Puddled clay

Slotted collector pipe in infiltration trench

Springbox (cover removed for clarity)

Apron slab

Drainage channel

Outlet, overflow and washout pipes

Self closing gate

Planted fibrous grass

Top soil

Water seepage zone

Puddled clay

Stone backfill

Collector pipe

Section A-A Infiltration trench

Figure 11 Spring with infiltration trenches and springbox.
Source: Guoth-Gumberger (1987)

provided with an overflow. The level of the overflow should be below the level of the lowest eye, so no water can flow back up any of the pipes towards the springs. A single pipe can deliver water from the inspection chamber to a water supply point, or to a storage tank, at a lower elevation.

If the flow from an aquifer is not concentrated, but is seeping out over a wide area, the water can still be intercepted. This can be done by constructing gravel-filled trenches across the slope, to collect the seeping water. Where appropriate, these trenches may contain perforated pipes to improve the rate of flow along them (Figure 11). Clay dams or impermeable walls should be built on the downhill side of each trench. These will maximize the amount of groundwater collected by preventing it flowing further downhill.

Sometimes a spring box is built at the eye of the spring. This is a small tank with a permeable wall through which the water can enter the tank (Figure 12).

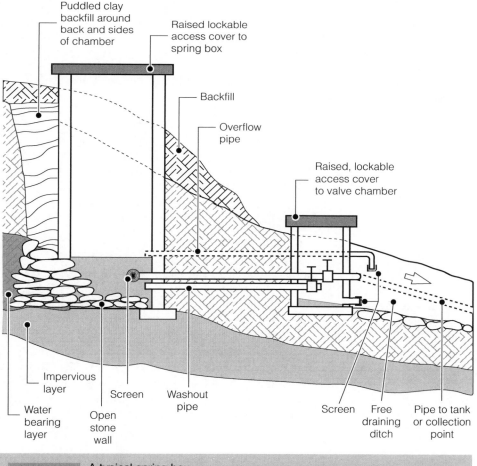

Figure 12 A typical spring box.

Whenever working near the eye it is important not to dig too far into the impermeable layer below the aquifer. If this layer is removed at any point the water from the aquifer may seep through the gap and be lost. This risk is greatest if a spring box is being constructed.

It should be noted that large spring boxes are often not necessary at the eye of a spring. If storage is needed then it is better if this is built some distance from, and below it. If this is done then only a shallow excavation will be needed at the eye to channel the water into a pipe leading to the tank. This shallow excavation avoids the risk of perforating the impermeable layer below the aquifer. Another reason to position the storage tank away from and below the level of the spring is to ensure that the level of its overflow is below the level of the eye.

BOX 9   An example showing the benefits of storage at springs

If a small spring is flowing at 2 l/min it would take 10 min for someone to fill a 20 l jerrycan, a time period that might frustrate any user. However over 24 hours this spring can provide (24 × 60 × 2) = 2880 l, enough for over 140 people to have at least 20 l per day. But, if there is no storage, to meet the needs of 140 people water would have to be collected at the spring throughout the day and night.

To make the best use of the water from the spring none must be wasted. It is necessary to calculate the volume of tank required to collect all the water which flows when it is not convenient for people to collect water. For example, if no one collects water between the hours of 7 p.m. and 6 a.m. (11 hours, or 660 minutes) the volume would need to be:

2 × 660 = 1320 l (in practice a greater volume should be provided to give a factor of safety).

Water could probably be delivered from the tank through a 20 mm diameter tap at about 20 l/min, as long as the tap was at least 8 m below the level of the base of the tank to give sufficient pressure to deliver the water this fast. A 20 l container could now be filled in 1 min instead of 10 min and the 140 people would only need to spend a total of 140 minutes (2.33 hours) of the day collecting water. However, since the tank's volume is 1320 l, when full, it only stores enough water for 66 people. This means that people would have to spread out their visits to the tank throughout the day, with sufficient gaps between some of them to allow the tank to partly refill so that no one completely empties it. If it did empty, then the rate at which the tap would deliver water would return to 2 l/min instead of 20 l/min.

As illustrated in Box 9, the main advantage of providing storage for spring water is that water can be delivered from the tank via a tap at a much faster rate than the rate at which it flows from the spring. Another important feature is that the tank will collect water which may otherwise be wasted (e.g. that which flows from a spring at night).

Note too, that a normal tap, fitted directly to the wall of a shallow tank, can only deliver water slowly because the water pressure is low. It is therefore much better if a tap is positioned well below a tank, connected to it by a pipe. Using this arrangement the water pressure will be greater, making the water flow from the tap at a faster and more convenient rate.

With springs it is important to prevent the water level in the aquifer rising above its seasonally highest level as this could cause the spring to move to another site. To avoid this, the overflow pipe from a spring box must always be positioned lower than the eye of the spring. Overflow pipes should be covered with mesh to stop amphibians and mosquitoes entering them and gaining access to the storage tank.

Any construction work at the spring eye should be covered with an impermeable layer such as 100 mm of concrete or 150 mm of puddled clay. This forms an impermeable layer suitable to stop soil particles or pollution easily entering the granular material placed at the point where the water leaves the aquifer (Figures 9 and 11). Puddled clay is formed by mixing clay soil with a little water and trampling it underfoot until it is uniformly flexible.

Before burying everything it is a good idea to record the position of the spring from some fixed point, or, after back-filling, to mark its position on the surface with a large stone. This will mean that the eye can be easily located, and re-excavated, should any problems occur.

The top of any spring box, inspection chamber or storage tank should be at least 300 mm above the local ground level, to prevent surface water running into it. Any access cover should be lockable and raised above the cover slab to exclude surface water.

Surface water should be diverted away from a spring to prevent the risk of it polluting the groundwater, which is very close to the surface at a spring. It can be diverted by creating a shallow ditch at least 8 m uphill and around each side of the spring (Figure 13). The soil from the ditch should be compacted in a ridge on the downhill side of the ditch as an added precaution. If a fence or prickly hedge is put on top of the ridge, this will help to keep people and animals away from the area immediately above the spring. The area inside the fence is best kept clear of bushes. It is also good to cover it with creeping grasses that are easy to maintain and that will protect the ground surface from erosion.

If, despite efforts to filter it, the water at the spring eye still carries a lot of suspended solids it may be necessary to provide a sedimentation chamber. Removing suspended solids is particularly important if the water is going to

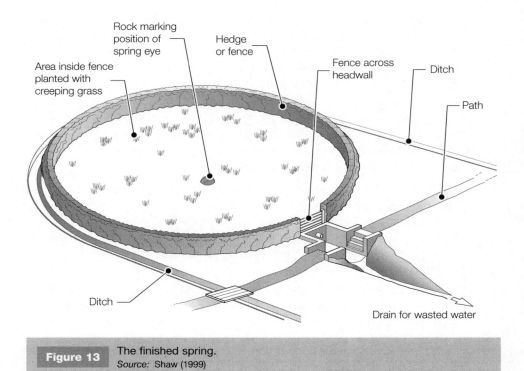

| **Figure 13** | The finished spring.
Source: Shaw (1999) |

be fed into a piped distribution system, which could become blocked by deposits. If no sedimentation tank is provided before a storage tank then the material will be deposited in the storage tank and may be re-suspended when the tank is nearly empty. A sedimentation chamber (Section 6.3) is ideally a long narrow tank (with a length of at least four times its width) or a wider tank with dividing walls which will ensure a long flow path from inlet to outlet. It should have a water depth of at least 1 m. To uniformly distribute the flow across the full width of the tank the inlet can be a perforated pipe, positioned horizontally across one end, at about mid-depth. The outlet should be a trough at the surface of the water. In normal operation the tank should remain full of water and should have sufficient volume to allow the water to flow across the tank slowly enough for most of the suspended materials to settle out. A volume equal to 30 to 60 minutes of flow will normally be sufficient. The floor should slope towards a washout pipe (Figure 56) fitted with a cap or a valve which can be opened to clean out the tank at least once a year, or more regularly if it is necessary to control the level of the settled silt so that it is not re-suspended.

Ideally new spring boxes and tanks should be disinfected before they are put into use as described in Section 3.5.

3.3.3 Boreholes and tubewells

Typical features of a borehole

The terms borehole and tubewell are often used interchangeably although strictly speaking a borehole is a hole which has been bored by a rotating bit, while a tubewell has been driven or excavated by jetting. In most of the following text the term borehole will be used to refer to both types of hole. Figure 14 shows the main features of a typical borehole.

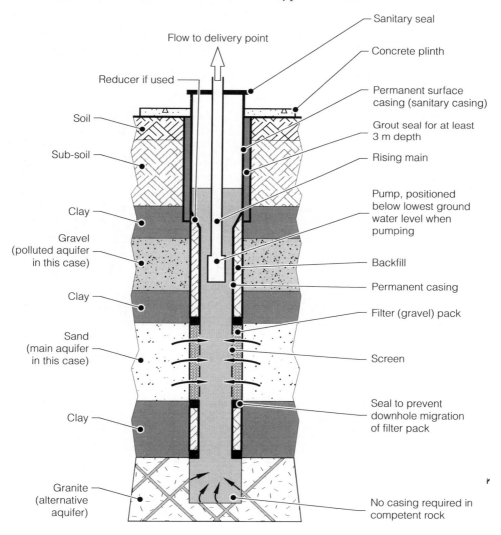

Note: Width of borehole is exaggerated for clarity

Figure 14 Schematic details of a typical borehole.
Source: WEDC

The purpose of a borehole is to allow the groundwater to be reached and extracted for water supply in a sustainable way. Among other things this means that any loose soil through which the borehole passes will need to be supported by a casing pipe which may be temporary or permanent. A permanent casing may be installed during drilling or may be installed inside a larger diameter temporary casing, which is used during drilling but which is subsequently withdrawn. To make it easier to withdraw the temporary casing its diameter can be progressively reduced as the hole deepens and this technique is known as telescoping casing. Each length of different diameter can then be removed separately. This has the advantage that each section will require less force to overcome the friction of the soil than if a single diameter of casing is used to line the whole of the borehole. However, this telescoping makes it necessary to use drilling bits of varying sizes. Nowadays, thick-walled plastic pipes are often used for permanent casings, but steel pipes may still be used for deep boreholes.

Where a borehole passes through firm rock it may be possible to dispense with casing. A driven tubewell uses the driven pipe as the permanent casing. When drilling in soils containing some clay the hole can sometimes be left temporarily unsupported until drilling is finished, at which stage the permanent casing is installed. For some methods of borehole construction drilling mud is used. This 'mud' is made from water thickened with a chemical additive or powdered clay, and may be circulated to remove excavated material and to lubricate and cool the cutting bit. This mud can carry away larger particles of excavated soil than is possible if only water is circulated. The drilling mud that temporarily fills the borehole can sometimes support the soil until drilling is finished, at which stage a permanent casing can be installed.

Below the groundwater level at least some of the casing needs to be slotted or perforated to allow water to enter the borehole. This section is known as the screen. The size and shape of the holes in the screen and the overall length of the screen pipe need to be chosen carefully. If they are too large there is a danger that soil from the aquifer may pass through them, causing the bottom of the borehole to eventually fill with sand or silt. However, if they are small enough to hold back all of the soil particles they will reduce the rate at which water can enter the borehole. This can be overcome by increasing the length of the screen, but this is expensive. Another solution is to use larger holes together with filter packs as described below. While simple screens can be made by using a hacksaw to cut a lot of short slots in the wall of a plastic pipe, purpose-made screens are much better. In granular soils, after construction is complete, the bottom of a borehole should be sealed to ensure that water enters the borehole only through the screen. This seal may be omitted if the base of the borehole enters rock.

To allow slots to be at least 1 mm wide an artificial filter pack (or gravel pack) is often installed in the gap between the inside of the borehole and the outside of the screen. This gap should be at least 50 mm wide. The pack is made up of a carefully graded mixture of small stones and very coarse sand and is poured into place as the temporary casing is withdrawn. Recently, as an alternative to the filter pack, there has been increased use of plastic meshes and textiles to form fine filters around the screen. The successful long-term performance of these filters has still to be proven.

Once the filter pack has been placed for the full length of the screen the top of the pack should be sealed to stop material from above entering the pack and blocking its pores (Figure 14). The remaining space above this point can then be filled with any suitable excavated material to a level about 3 m below the ground surface.

Near the surface the top 3 m of the gap between the temporary and permanent casing should be filled with impermeable material. This is to form a sanitary seal that prevents any pollution from the surface easily passing down the gap to contaminate the groundwater. Compacted clay is one suitable material. Cement grout, a thick mixture of cement and water, is also suitable, and more convenient.

Once a borehole has been constructed, it is often advantageous to develop it to improve its yield. This moves the finer particles in the aquifer away from the filter pack so that the water can flow towards the screen more easily. Development also removes any drilling mud that remains in the pores of the aquifer.

Development is accomplished either by pumping intermittently from the borehole at a very high rate, or by using a surge plunger in the borehole casing. The surge plunger, which is rather like the piston of a handpump, is rapidly raised and lowered in the permanent casing causing groundwater in the pack to pass rapidly to and fro. During this process, smaller particles of soil are washed into the borehole. The water movement in and out of the slots sets up a natural grading of particles with the larger sizes nearest to the slots and the finer further away. This forms a graded filter, both in the pack and in the adjacent part of the aquifer. When the borehole is in use this filter prevents the ingress of fine material from the aquifer while still providing a fairly easy flow path for the groundwater to enter the borehole via the screen. Once the borehole has been developed its yield can be tested (see Section 3.6).

The details about filter packs and borehole development given above relate to the ideal situation. Some practitioners are of the opinion that borehole development and use of a filter pack are often unnecessary if a borehole is to supply only a relatively low pumping rate, such as that needed by a typical handpump. Obviously with a driven tubewell no artificial filter pack can be installed, although in a well-graded aquifer development can be used to produce a natural filter pack around the screen.

The way in which spilt and other wasted water is to be disposed of should be carefully considered when siting a borehole. For example, if at all possible, boreholes should not be sited where it will be difficult to drain the spilt water, or surface water, away from the borehole. A sloping concrete apron slab, about 2 m diameter, or 2 m × 2 m square, with a raised edge, should be constructed around the borehole to prevent muddy conditions developing if water is spilled. This slab will also reduce the risk of polluted water seeping down the edge of the borehole casing (although where a temporary casing was used the sanitary seal around the permanent casing should prevent this happening). The apron slab should be connected to a drainage channel to carry water away from the slab to a suitable disposal point (e.g. a drainage ditch, a soakaway pit or soakaway trench, a garden, etc.). The slab and channel should be constructed on firm ground. Loose soil should be compacted or excavated and replaced with a well-compacted layer of rocks and stones. Normally the apron slab is made from reinforced concrete, typically a 100 mm thick slab with 8 mm bars at 150 mm spacing in both directions. The lowest bars should be positioned 50 mm above the base of the slab.

Ideally the borehole casing should rise above the surface of the slab and fit inside the pumphead (Figure 44). This will prevent any water which passes under the bottom edge of the pumpstand from reaching the top of the casing and flowing into the borehole.

Hand-augered boreholes

In granular soils boreholes can be constructed using hand tools. Most tools are connected to a strong steel pipe that is rotated manually to excavate the soil. These tools are called augers.

Another method of excavation is to repeatedly raise and drop a heavy chisel bit or hollow cylindrical bit so it cuts away the ground. This cable-tool or percussion method was first used many hundreds of years ago. Nowadays it is rarely hand-operated but instead relies on very heavy tools lifted by a motorized winch.

Although the auger tools can be made in a workshop they are usually purchased from a manufacturer. Different designs of tool are needed depending on the nature of the soil and whether the drilling is taking place above or below the water table. Some tools have a spiral to hold the excavated soil; others are designed so that the soil sticks to the inside of a cylinder or is held inside it by a hinged flap, or internally projecting teeth (Figure 15). Tools that are used below the water table often have valves to hold the water inside them as well as the excavated soil.

During use of an auger the tool has to be lifted out of the hole periodically, to remove the soil which has been loosened. As the depth increases, additional sections of drilling rod need to be added. Normally, to remove the excavated soil, everything has to be lifted out after every 0.5 m of progress. After the tool is cleaned everything then needs to be installed again. This can be a very tiring and tedious process, particularly with deep boreholes. A manually operated winch may be used to safely raise the heavy weight of the drilling rods, drilling bit and excavated soil (Figure 15). To improve the rate

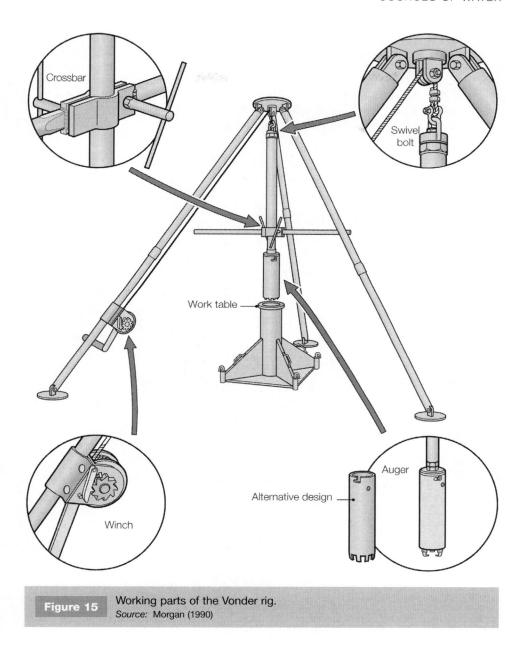

Figure 15 Working parts of the Vonder rig.
Source: Morgan (1990)

at which the drilling tool cuts into the soil people may be asked to sit on a crossbar connected to, and be rotated with the drilling rod.

Temporary and permanent casings, screen and filter pack might all be necessary with a hand-augered borehole.

As an alternative to lifting out the excavated soil on or in the drilling/excavating tool, one manual drilling method circulates a thick drilling mud to carry the loosened soil out of the hole. This method is only suitable in fine soils. The mud can be pumped manually using a foot-powered pump (Figure 16).

There is insufficient space in this short book to fully discuss the procedures followed during hand-augering. More information may be found in the sources recommended in the appendices and references.

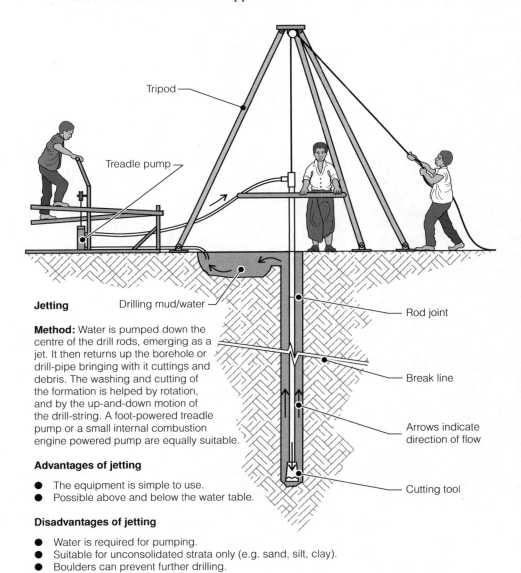

Tripod

Treadle pump

Jetting

Drilling mud/water

Method: Water is pumped down the centre of the drill rods, emerging as a jet. It then returns up the borehole or drill-pipe bringing with it cuttings and debris. The washing and cutting of the formation is helped by rotation, and by the up-and-down motion of the drill-string. A foot-powered treadle pump or a small internal combustion engine powered pump are equally suitable.

Rod joint

Break line

Arrows indicate direction of flow

Cutting tool

Advantages of jetting

● The equipment is simple to use.
● Possible above and below the water table.

Disadvantages of jetting

● Water is required for pumping.
● Suitable for unconsolidated strata only (e.g. sand, silt, clay).
● Boulders can prevent further drilling.

Figure 16	Jetting.
	Source: Shaw (1999)

Machine-drilled boreholes

Unlike methods of hand-drilling, motorized drilling rigs are able to cope with all kinds of soil/rock conditions at virtually any depth. There are two basic kinds of machine, the percussion rig and the rotary rig.

The simplest form of rig is a cable tool or percussion rig, which drills the hole by using a wire rope and winch to repeatedly raise and drop a heavy bit at the bottom of a hole. A shell bit, which consists of a hollow cylinder with a cutting edge, is used to excavate soils that contain clay. For granular soils which do not have clay to bind them together a valve is fitted to the bottom of the cylinder to hold any water and excavated soil inside the tool. This cylinder is called a bailer. To break up rock a heavy chisel bit is used. After the rock has been broken up into small particles by the chisel bit, if necessary, water is added to the hole to form a slurry. The chisel is then replaced by the bailer, which is used to remove the excavated material from the hole.

The second type of drilling rig is a rotary rig that uses a heavy rotating bit to break up the soil/rock in the hole. It can drill holes much faster than percussion rigs and with suitable bits can drill holes in very hard rock. Such rigs need equipment for rotating, lifting and clamping the drill pipe. They are much more expensive than percussion rigs. They also require a soil removal system, either pumps to circulate water (or drilling mud, see p. 34), or a powerful air compressor to circulate large volumes of air (sometimes with drilling foam).

Down-the-hole hammer (DTH) drilling is a mixture of percussion and rotary drilling which is particularly useful in hard rocks. A heavy air-actuated single piston hammer positioned at the bottom of a drill shaft provides the percussion. The bit is rotated to ensure that the hardened teeth on the bit move across the whole surface at the bottom of the hole. The broken rock particles are usually flushed out by the air, which also cools the bit.

Poor access to sites in rural areas means that motorized rigs, which are usually quite heavy, are often unable to travel and work during the rainy season. Their weight may also mean that even in the dry season they need a compacted gravel track to be constructed from the nearest road to the borehole site. Fortunately in recent years much smaller drilling rigs, able to drill boreholes suitable for handpumps (i.e. to depths usually less than 40 m and with permanent casing diameters of 125 mm or less) have become available (Figure 17). These have considerably increased the number of potential sites that can be accessed. Many of these small rigs use a mixture of human and engine power (via shafts and gears) which makes them much simpler than conventional hydraulically powered rigs. This means that they are easier to maintain and repair, which reduces the cost of the boreholes they drill. Some rigs are very portable, capable of being carried in the back of a pick-up truck and then being reconstructed from pieces which can be manhandled into place. Most of these rigs use the rotary drilling method but a few have DTH bits.

Method: A drill-pipe and bit are rotated to cut the rock. Air, water, or drilling mud is pumped down the drill-pipe to flush out the debris. The velocity of the flush in the borehole annulus must be sufficient to lift the excavated debris.

Advantages of rotary drilling (with flush):

● Most rock formations can be drilled.
● Water and mud supports unstable formations.
● Fast.
● Operation is possible above and below the water table.
● Possible to drill to depths of over 40 metres.
● Possible to use compressed air flush.

Disadvantages of rotary drilling (with flush):

● Requires capital expenditure on equipment.
● Water is required for pumping.
● There can be problems with boulders.
● Rig requires careful operation. and maintenance.

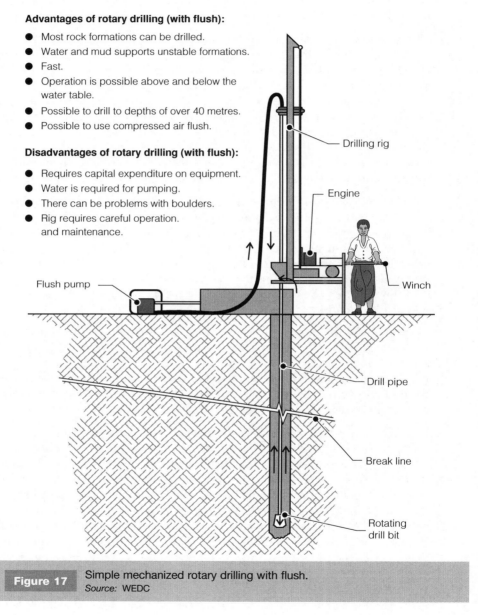

| Figure 17 | Simple mechanized rotary drilling with flush. |
Source: WEDC

Drilling can be very expensive, so expert advice about the probability of finding suitable water at a proposed site should be obtained before arranging for drilling to take place. The competence of any drilling contractor should be investigated before employing him. Ideally he should be supervised to ensure that he does not take shortcuts which may compromise the future performance of the borehole (e.g. by not properly installing a sanitary seal or filter pack, or not drilling to the required depth in the aquifer).

Driven tubewells

A driven well is made by pushing a strainer called a wellpoint into the ground. A wellpoint is a specially made metal tube with a point at the lower end, and holes or slits in the sides through which water can filter into the pipe. These holes are made in special ways, to prevent them becoming clogged with soil (e.g. in a wedge-wire screen the slot is wider on the inside than on the outside). Home-made wellpoints are not as strong or as durable as manufactured ones and the slots made in them are easily blocked during installation. The supplier of a manufactured wellpoint should be asked for detailed advice on how to install it.

Usually the wellpoint is driven into the ground by hitting the top of the pipe with a heavy weight suspended from a tripod. A drive head is fitted to the top of the pipe to protect it from damage and a guide rod may be used to keep the driving weight central. (Figure 18). With some wellpoints a metal bar is temporally installed inside the pipe to transmit the force directly to the bottom end of the pipe. As a wellpoint is driven further into the ground, additional lengths of steel pipe are screwed on at the top. This string of pipes acts as the rising main pipe for the water that will be pumped from the well when it is completed. To make sure that the pipe joints do not unscrew, the pipe should be twisted clockwise after each blow of the hammer.

Most wellpoints are only 30–50 mm diameter, which is too small to allow usual handpump cylinders to be installed inside them. Driven wells are therefore best suited to suction pumps, where the cylinder can be positioned at the surface. This arrangement will mean that water cannot be lifted from more than about 7.5 m below ground level (see comments on the suction limit in Section 4.5.2).

Occasionally it is desirable to first construct a borehole to the appropriate depth using one method and then to drive the wellpoint several metres into the aquifer below the base of the borehole. Driving the wellpoint into the ground at the bottom of a borehole reduces the length that has to be driven, and thus the required driving force. However, it is only in some special situations that use of both a borehole and a wellpoint would be more appropriate than providing a screened section of casing in the borehole. Where a wellpoint is installed in the bottom of a borehole, a deepwell pump with a diameter larger than the wellpoint can be positioned in the larger diameter borehole, as long as the cylinder will always be below groundwater level.

An artificial filter pack cannot be used with a wellpoint, but, in suitable soils, proper development may form a natural filter pack which can improve the yield of the wellpoint.

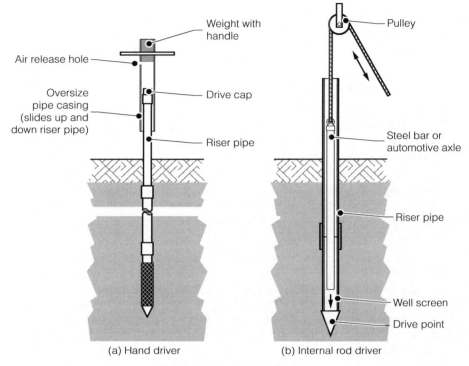

Air release hole

Oversize pipe casing (slides up and down riser pipe)

Weight with handle

Drive cap

Riser pipe

Pulley

Steel bar or automotive axle

Riser pipe

Well screen

Drive point

(a) Hand driver

(b) Internal rod driver

Methods for driving well points

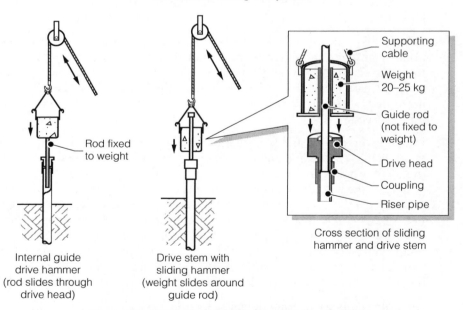

Rod fixed to weight

Internal guide drive hammer (rod slides through drive head)

Drive stem with sliding hammer (weight slides around guide rod)

Supporting cable

Weight 20–25 kg

Guide rod (not fixed to weight)

Drive head

Coupling

Riser pipe

Cross section of sliding hammer and drive stem

Heavy duty sliding hammer and drive stem assemblies

Figure 18 Methods for driving wellpoints.

Jetted tubewells

As mentioned in Section 3.3.1 a jetted tubewell is formed using pumped water to excavate a narrow hole into the soil. The water is pumped down a pipe that is pushed down into the ground. The fast flowing water exiting from the bottom end of this pipe washes away the soil. The loosened soil particles are flushed up the hole that has formed around the pipe. This excavation process can be assisted by raising and dropping the pipe and/or rotating it so that a bit on the bottom end of the pipe loosens the soil so it can be carried away by the water.

Jetting normally requires plenty of water, although a re-circulation pit can be dug at the surface so that the heavier solids can settle out and the raised water can be reused. It usually needs a motorized pump, and various special fittings, but a manually operated treadle foot pump (Figure 16) has been successfully used. Sometimes additives (such as bentonite clay) are used with the water to make it thicker to form a 'drilling mud'. This improves its ability to carry heavier particles of soil and reduces the amount of water that escapes through the sides of the hole during construction. In suitable soils this method can be used to sink wells up to 80 m deep.

When the hole has reached the desired depth the jetting tool can be removed and a permanent casing with a screen can be installed in the hole as for a normal borehole.

It is possible to jet the screen and permanent casing directly into the ground. However, unless a special type of screen is used much of the jetting water will leave through the sides of the screen, reducing the rate of excavation. Some manufacturers therefore produce a special screen containing a valve which causes all the jetting water to exit from its bottom end during jetting. When this screen is jetted into the ground all the pumped water contributes to excavating the soil below the bottom of the screen. When the screen reaches the right depth jetting stops. As soon as water is extracted from the casing, the valve at the bottom of the screen closes, so that all the groundwater entering the casing passes through the screen.

Sometimes jetting can be carried out using two pipes at the same time, one being the permanent casing (and screen) and the other a pipe which is only used for jetting.

It is advantageous in very fine soils if the particle sizes in the aquifer around the screen are altered by adding coarse sands and fine gravel to the hole. These should be added just before finishing jetting, but the pumping rate should be reduced so that these coarse particles can sink to the bottom of the borehole.

The sludger method of forming a tubewell

The sludger or sludging method is a simple method of reverse jetting that needs no pump. It is only useful in fine loose soils, such as sands and silts, and becomes more difficult as the depth to the groundwater increases. It is most appropriate for delta areas or flood plains where the soil is fine silty sand and the groundwater is usually within 10 m of the surface

To construct a tubewell using the sludger method, a small hole about a metre deep, dug by hand where the well is to be sited, is filled with water. A larger shallower hole is usually excavated to one side of it to capture the water discharged during the sludging process. This second pit allows heavier solids to settled out before the water is recirculated. Then a piece of steel pipe, about 50 mm diameter and 3 m or more in length, is inserted into the hole. A bit, such as a pipe coupling sharpened at the edge, is fixed to the bottom of the pipe to help it cut its way into the soil. A scaffolding of wood or bamboo is built beside the hole, and a lever fixed to it, with one end tied to the pipe by a chain (Figure 19), so that the pipe can be lifted up and down in the hole by operating the lever. A man stands on the scaffolding and uses his hand or a flat piece of rubber as a valve. As the pipe rises, he holds his hand (or the rubber) firmly over the end so that any water and soil in the pipe is lifted. As it falls he lets go allowing more soil and water to enter the bottom of the pipe, displacing what is already in the pipe and causing it to be discharged from the top end. Repeating this up-and-down process pumps water up the pipe, bringing some of the soil with it. As the hole deepens the pipe can be advanced lower and lower into the ground. When necessary, further lengths of pipe should be added at the top. A 6 m length of temporary casing, about 20 mm larger in diameter than the pipe, may be installed to support the top of the hole when it reaches that depth. Other than this piece of casing at the top, the hole usually remains un-cased until the hole is complete. At this stage the sludger pipe is pulled out carefully, keeping the hole full of water or drilling mud. A permanent casing and screen can then be inserted.

The hole should be kept full of water during sludging. If the soil is very permeable and the water table is deep, it may be difficult to keep the hole full. Drilling mud can be used to reduce the rate of seepage into the soil above the groundwater level. Traditionally in Bangladesh a little cow dung is added to the water (about one part in twenty) to help to seal the soil and slow down the rate at which the water seeps out of the hole. The dung need not be dangerous to health as long as the well is properly flushed out and disinfected on completion (Section 3.5). Ideally drilling mud should only be used during excavation of the top part of the hole. If it is used to support the hole around where the screen will later be positioned it may block the pores and slow down the rate at which water can seep into the well when it is completed. However, good development after the permanent casing is installed, and use of a biodegradable drilling mud should minimize such problems.

The hole around the installed casing should be backfilled and sealed.

Sludging (reverse jetting)

Method: This method has been developed and used extensively in Bangladesh. A hollow pipe of bamboo or steel is moved up and down in the borehole while a one-way valve - a hand can be used to improvise successfully - provides a pumping action. Water flows down the borehole annulus (ring) and back up the drill pipe, bringing debris with it. A small reservoir is needed at the top of the borehole for recirculation. Simple teeth at the bottom of the drill pipe, preferably made of metal, help cutting efficiency.

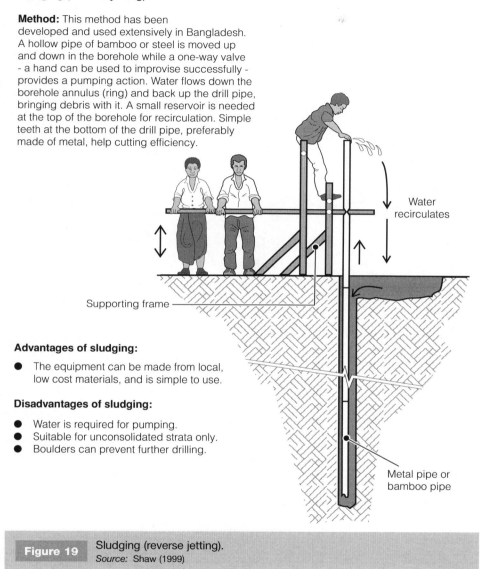

Water recirculates

Supporting frame

Advantages of sludging:

● The equipment can be made from local, low cost materials, and is simple to use.

Disadvantages of sludging:

● Water is required for pumping.
● Suitable for unconsolidated strata only.
● Boulders can prevent further drilling.

Metal pipe or bamboo pipe

Figure 19	Sludging (reverse jetting). *Source:* Shaw (1999)

3.3.4 Hand-dug wells

Digging a well by hand can be dangerous for the workmen so it is advisable to have an experienced supervisor in charge of the work. The process is too complicated to describe adequately in this small book. Appropriate guidance is available from a number of sources mentioned in the appendices and references. Risks to workers include collapse of the excavation, items being dropped down the well and suffocation by dangerous gases from the soil or

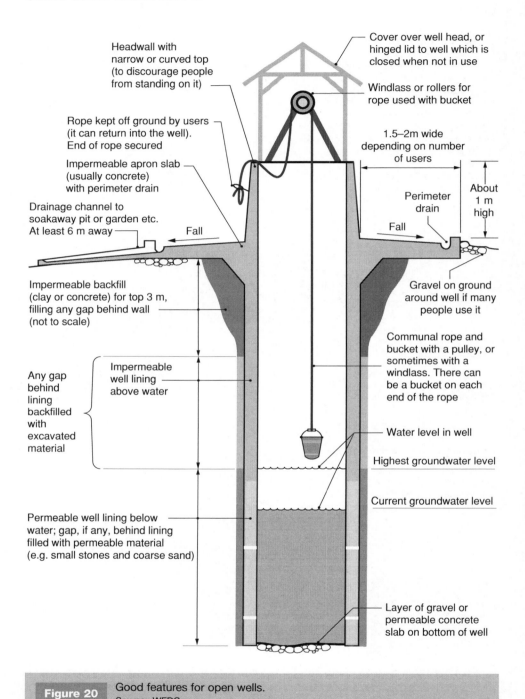

Headwall with narrow or curved top (to discourage people from standing on it)

Cover over well head, or hinged lid to well which is closed when not in use

Windlass or rollers for rope used with bucket

Rope kept off ground by users (it can return into the well). End of rope secured

1.5–2m wide depending on number of users

Impermeable apron slab (usually concrete) with perimeter drain

Perimeter drain

About 1 m high

Drainage channel to soakaway pit or garden etc. At least 6 m away

Fall

Fall

Impermeable backfill (clay or concrete) for top 3 m, filling any gap behind wall (not to scale)

Gravel on ground around well if many people use it

Communal rope and bucket with a pulley, or sometimes with a windlass. There can be a bucket on each end of the rope

Any gap behind lining backfilled with excavated material

Impermeable well lining above water

Water level in well

Highest groundwater level

Current groundwater level

Permeable well lining below water; gap, if any, behind lining filled with permeable material (e.g. small stones and coarse sand)

Layer of gravel or permeable concrete slab on bottom of well

Figure 20 Good features for open wells.
Source: WEDC

internal combustion engines. A petrol or diesel engine should never be used in, or close to, the well. This is because odourless gases from the engine's exhaust pipe can enter and kill people in the well.

Where the soil is suitable, producing a hand-augered hole on the proposed site of the well can help establish the types of stratum that will need to be excavated and the level at which groundwater will be reached. With suitable equipment the safe rate at which water can be drawn from the borehole can also be checked to give some idea about the behaviour of the aquifer. If the site is found to be unsuitable the costs of constructing the well will have been avoided.

The good features of a completed open well are illustrated in Figure 20. The headwall, apron slab, drainage channel and seal (3 m of impermeable backfill) are used to reduce the risk of surface water and spilt water easily entering and contaminating the well. A cover to the well is advisable.

BOX 10   The main sections of a hand-dug well

There are three essential sections to a well:

- the headworks at the surface. This is designed to reduce the potential for contamination of the well and to make it as easy as possible for people to collect water from it

- an impermeable lined shaft to support the soil, and to prevent polluted surface water from seeping into the well

- an intake section, which can be an extension of the shaft. This is designed to hold back the soil but to allow water to enter the well. It is therefore strong, but porous or perforated. To stop the soil on the base of the well being disturbed it should be covered with a thick layer of gravel or pre-cast segments of a circular concrete slab. Sufficient storage volume to suit the demand pattern needs to be provided in the well. All of this storage volume needs to be below the lowest annual groundwater level.

Figure 21 shows various arrangements of shaft and intake which can be used for different strengths of soil.

Box 10 explains the purpose of the three main sections of a hand-dug well. As illustrated in Figure 21 the need for, and method used to construct the shaft will vary with the strength of the soil. The shaft is usually constructed from concrete or brickwork. This lining may be periodically built up from the base of the excavation as it is deepened. The amount of excavation that can safely be left unsupported before a section of the shaft is constructed needs experienced judgement. Sometimes no more than 0.5 m may be safely left unsupported. Where the soil is prone to collapse, the caissoning method is the safest construction technique. In this method all excavation takes place

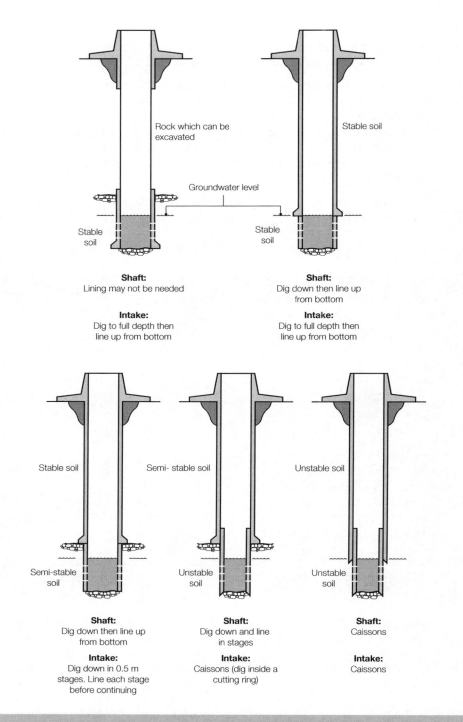

Rock which can be excavated

Stable soil

Groundwater level

Stable soil

Stable soil

Shaft:
Lining may not be needed

Intake:
Dig to full depth then line up from bottom

Shaft:
Dig down then line up from bottom

Intake:
Dig to full depth then line up from bottom

Stable soil

Semi- stable soil

Unstable soil

Semi-stable soil

Unstable soil

Unstable soil

Shaft:
Dig down then line up from bottom

Intake:
Dig down in 0.5 m stages. Line each stage before continuing

Shaft:
Dig down and line in stages

Intake:
Caissons (dig inside a cutting ring)

Shaft:
Caissons

Intake:
Caissons

Figure 21 Suggested digging and lining techniques for hand-dug wells in different strata.
Source: WEDC

inside a lining which sinks into the ground as the soil is removed from inside. A stack of pre-cast concrete rings is a suitable lining for this method.

Recently the modified Chicago method (Figure 22) of temporary support for well excavations has been introduced in some countries. The vertical planks of wood used with this method can be closely spaced in fine granular soils, and at larger centres in stronger soils. The planks can be gradually withdrawn as a permanent lining is built up from the bottom of the excavation. Should insufficient water be found, then the planks can be progressively removed and the well can be refilled and abandoned without loss of any construction materials. However, if the well had been constructed using a more conventional method then the permanent lining materials would have been wasted if insufficient water was found.

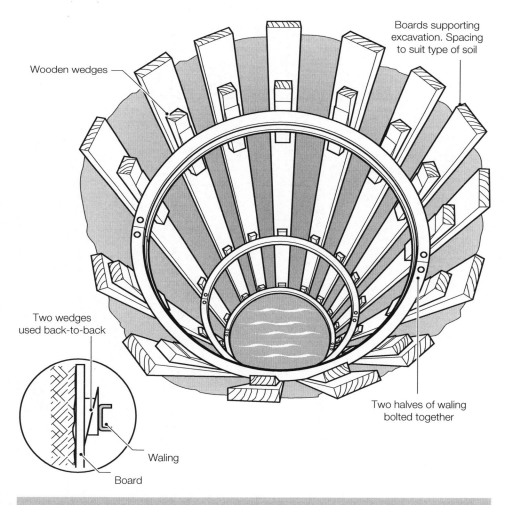

Boards supporting excavation. Spacing to suit type of soil

Wooden wedges

Two wedges used back-to-back

Two halves of waling bolted together

Waling

Board

Figure 22	Well excavation supported by the modified Chicago method.
	Source: WEDC

In some situations, to reduce the cost of the well it may be appropriate to backfill the part which is above groundwater level (Figure 23) or to backfill around a vertical pipe in the centre of the well (Figure 24). However, both of these methods of finishing the well would prevent it being deepened easily should the level of the water table fall. They would also necessitate the use of a handpump (or a Blair bucket pump, see Section 4.2.1) for collecting the water.

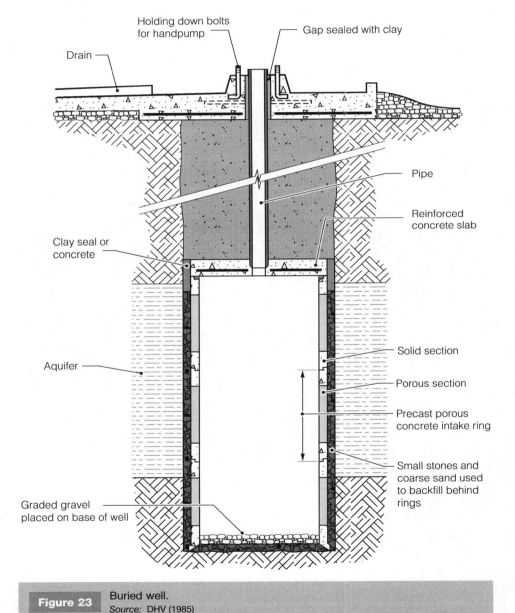

Figure 23	Buried well.
	Source: DHV (1985)

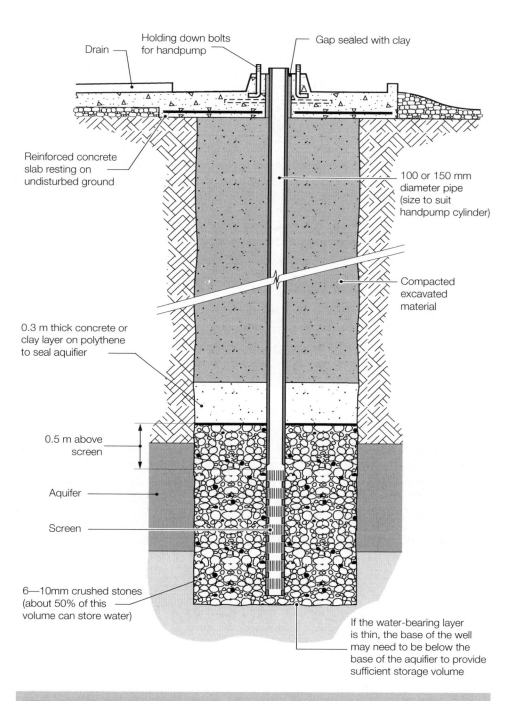

Drain

Holding down bolts
for handpump

Gap sealed with clay

Reinforced concrete
slab resting on
undisturbed ground

100 or 150 mm
diameter pipe
(size to suit
handpump cylinder)

Compacted
excavated
material

0.3 m thick concrete or
clay layer on polythene
to seal aquifer

0.5 m above
screen

Aquifer

Screen

6—10mm crushed stones
(about 50% of this
volume can store water)

If the water-bearing layer
is thin, the base of the well
may need to be below the
base of the aquifer to provide
sufficient storage volume

Figure 24   Backfilled well.
Source: DHV (1985)

It may be appropriate to improve an existing open well rather than dig a new one. Figure 25 shows one way of doing this. Use of a handpump is a much more hygienic way of collecting water than using a bucket and rope. However, the community must be willing to accept the pump and be able to maintain it. If a handpump is not acceptable then a bucket and windlass, or a shaduf (Section 4.2.1) are other options which can keep the bucket and rope off the apron slab where they can pick up pollutants left from people's feet. Note that if a handpump, shaduf or windlass is installed, only one person can draw water at a time, whereas traditionally at an open well it may be possible for several people to collect water at the same time if they use separate buckets and ropes. If buckets must continue to be used it will be much easier for people to collect water if roller bars (Figure 34) are positioned across an open well, or a pulley is provided. This is because these devices make it unnecessary for people to lean over the well throughout the lifting process. They also make the rope less likely to wear out and help prevent the bucket being damaged by hitting the wall of the well.

If new wells are sited near to existing ones that cannot be improved the latter should be carefully backfilled, ideally with 3 m depth of clay at the surface. This needs the agreement of users of the existing well. If an existing well is not filled in this way it can pollute the groundwater which will be collected by the new well.

Periodically, open wells will need cleaning. The best time for this is at the end of the dry season when the groundwater level is lowest. A powerful pump, like the one needed to de-water the well during construction, may be needed if the well cannot be emptied fast enough with a large bucket. The inside of a well can be inspected from the surface, using a shaft of light reflected from a mirror.

Before anyone enters a well a lighted candle should be lowered to test for harmful gases. If it flickers, or goes out, the well should be ventilated before anyone enters. Ideally this should be carried out using a piped ventilation system which feeds air to the bottom of the well using manually operated bellows or wind pressure. If this arrangement is not feasible an alternative is to use a long rope to rapidly raise and lower a large bunch of leafy twigs throughout the depth of the well a number of times. After this the candle test should be repeated.

During both construction and maintenance, dangerous gases may enter the well, while workers are inside. Therefore, each person should be properly secured to a safety rope tied round their upper body in such a way that it cannot slip off if they become unconscious. The rope should pass over a secure pulley on a strong frame positioned above the well, and a team of people should stand by to pull each person out quickly if necessary.

After cleaning, a well should be disinfected (see Section 3.5).

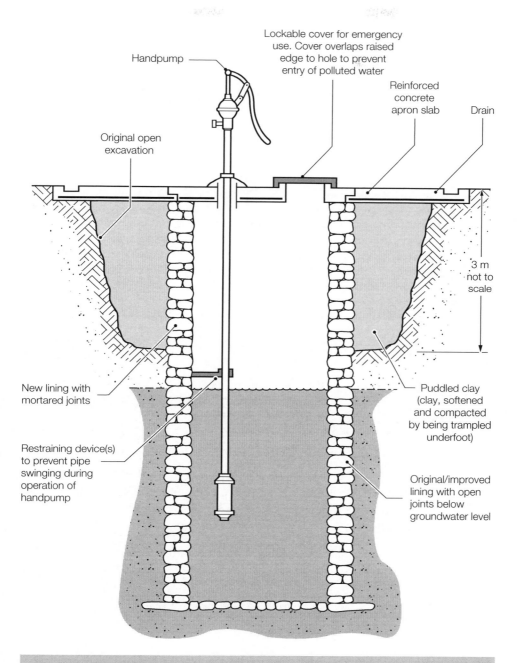

Handpump

Lockable cover for emergency use. Cover overlaps raised edge to hole to prevent entry of polluted water

Reinforced concrete apron slab

Drain

Original open excavation

3 m not to scale

New lining with mortared joints

Puddled clay (clay, softened and compacted by being trampled underfoot)

Restraining device(s) to prevent pipe swinging during operation of handpump

Original/improved lining with open joints below groundwater level

Figure 25   Cross-section of upgraded well with handpump.
Source: WEDC

3.4 Surface water

3.4.1 Introduction

As mentioned in Box 2, because of its poor quality surface water should be avoided for potable purposes unless it can first be treated. The only exception to this rule might be where water is being taken from a mountain stream which is flowing from a catchment area in which no significant human or animal activity takes place, but even this is not without risks.

Some surface water pollution is naturally filtered out as the water passes through permeable banks or beds of streams. The following section will first examine the use of infiltration systems for collecting such filtered surface water, but it can also be extracted using the wells and boreholes discussed in Section 3.3. Where the river is seasonal and the suspended solids it carries are mainly sands and gravels, sand dams are another way of collecting filtered surface water. The construction of a sand dam, or any other kind of infiltration system, involves a great deal of work. An engineer should therefore be asked to advise you about appropriate design and construction. Box 11 summarizes some important points about surface water intake.

BOX 11   Introduction to surface water intakes

- Direct extraction of surface water should be avoided unless further treatment is being provided. Groundwater extracted from near to surface water sources will be of much better quality than the surface water so if no treatment can be provided choose this option wherever it is feasible.

- The design of raw surface water intakes is best done by engineers. Such intakes, particularly those on rivers, often have to cope with:

  - fluctuating water levels and sudden changes in velocity of flow

  - a variety of different sizes of solids carried in the water, including large floating objects which, during floods, may damage the intake

  - possible movement of the river or stream away from the site of the intake.

- Intakes can be categorized into:
  - infiltration systems
  - above bed intakes: fixed or floating
  - below-bed intakes: Tyrolean weirs.

Section 3.4.4 presents appropriate design features for surface water intakes to collect water for treatment. Chapter 6 briefly examines the options available for partially or fully treating surface water. These will include sand filtration, a similar, but not identical, process to the infiltration systems described below.

3.4.2 Surface water infiltration

Infiltration galleries

Near to permanent sources of surface water, such as a stream, lake or village pond, the groundwater table can be expected to be very near the surface (Figure 1). It is preferable to use this groundwater rather than to use the surface water directly. However, in some areas the groundwater may not be of potable quality, or the ground conditions may not allow it to be easily abstracted.

The main difficulty with constructing an infiltration gallery is that it is necessary to dig a trench for it more than 1 m below the groundwater level. This normally will require continuous pumping with powerful pumps to keep the trench dry, and the sides of the trench will also need to be temporarily supported.

If the soil around a surface water source is not permeable enough for sufficient water to be extracted from a well, it may be possible to build an infiltration gallery in the bed of the body of surface water. This is basically a horizontal tubewell. However, instead of a casing, a perforated collector pipe is laid in gravel and sand so that water can drain into it. These materials must be carefully placed around the pipe in layers, ideally, so that the layers become finer the further they are away from the pipe. This, like the filter pack mentioned earlier, prevents particles flowing into the pipe and blocking it. Where the bed of the pond is impermeable simple horizontal beds of different granular materials can be used (see Section B-B on Figure 26).

Where the bed is permeable, ideally the layers should surround the pipe. Laying the separate vertical layers on each side of the pipe is not easy. However, during the placing of the granular material roofing sheets can be used vertically, to form temporary divisions between the two or three different sizes of granular material which are backfilled on each side of the pipe.

Wrapping the pipe, or the coarse gravel that surrounds it, in an appropriate civil engineering 'geotextile' is an alternative method of excluding fine material from the pipe.

The collector pipe should be closed at one end. The other end should lead the water into a collector well, from which it can be pumped in the usual way. Water can also be pumped directly from the gallery, although this is not usually advisable as it may lead to a lower quality of water (see Section 3.4.3).

One way to avoid the problem of constructing an infiltration gallery below groundwater is to build the gallery in dry ground to one side of the stream or pond. The excavation above the filter then forms a channel or pond into which some of the water from the source is diverted (Figure 27). Ideally the inlet connection from the main water source should be designed so that when necessary the flow of water from the source can be stopped. Also a gravity flow outlet from the pond/channel should be provided to allow water

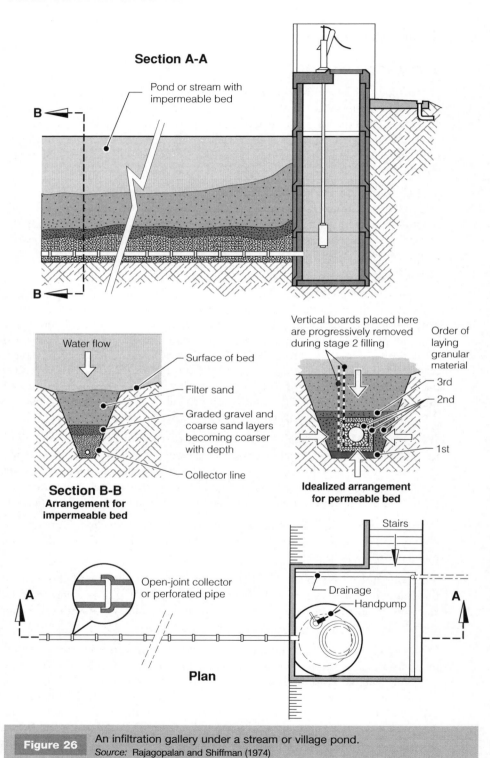

Section A-A

Pond or stream with impermeable bed

B

B

Water flow

Surface of bed

Filter sand

Graded gravel and coarse sand layers becoming coarser with depth

Collector line

Section B-B
Arrangement for impermeable bed

Vertical boards placed here are progressively removed during stage 2 filling

Order of laying granular material

3rd

2nd

1st

Idealized arrangement for permeable bed

Stairs

Open-joint collector or perforated pipe

Drainage

Handpump

A

A

Plan

Figure 26	An infiltration gallery under a stream or village pond.
	Source: Rajagopalan and Shiffman (1974)

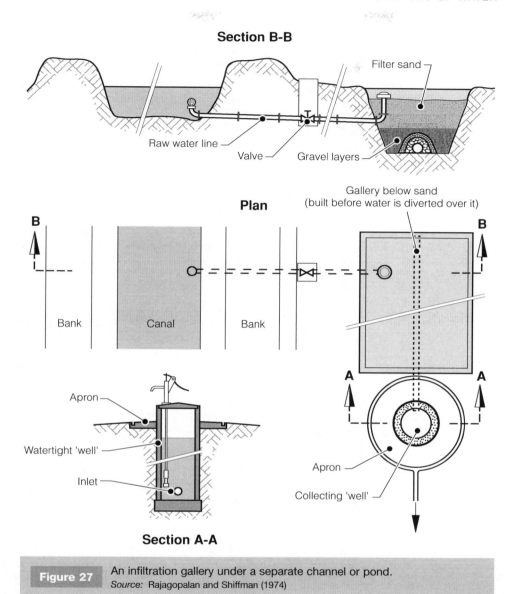

Section B-B

Filter sand

Raw water line

Valve

Gravel layers

Gallery below sand
(built before water is diverted over it)

Plan

B B

Bank Canal Bank

Apron

Watertight 'well'

Inlet

Apron

Collecting 'well'

A A

Section A-A

Figure 27 An infiltration gallery under a separate channel or pond.
Source: Rajagopalan and Shiffman (1974)

above the bed to be drained off when necessary. With this arrangement it will be possible to periodically drain the water from the channel/pond. This will expose the surface of the filter so that deposits that are blocking the pores on its surface can be cleaned off. While the surface of the bed is exposed, water can be pumped into the bed through a pipe pushed into the sand to wash out deeper deposits of fine material trapped in the filter. If at the same time a small flow of water from the inlet is allowed to flow over the bed it will carry away the deposits which have been disturbed.

Another option is to construct a gallery above, or not very deep below, the groundwater level and then after construction to make the groundwater level rise. For example, a gallery can be constructed in the bank of a river, followed by construction of a dam to raise the water level in the river. This will raise the local groundwater level and also has the additional advantage that water stored behind the dam will replenish the groundwater. Another option is to construct the gallery in the bed of a seasonal river and then to construct a dam across the river. Such dams can have problems with fine solids settling out of suspension behind the dam. These deposits will restrict the flow of groundwater into the bed and therefore into the infiltration system. This siltation is less of a problem if periodically water is flowing in the river with sufficient velocity to resuspend and carry away the deposits.

When an infiltration system is positioned in the bed of river there is a risk that during floods the bed may be eroded, exposing the gallery to damage. It therefore needs to be constructed deep enough to avoid this risk.

Sand dams and groundwater dams

In some situations, particularly with ephemeral streams in arid areas, the solids carried in the water are mostly sands and gravels. If a dam is built the volume behind the dam can be allowed to fill up with these deposits. This forms a **sand dam** (Figure 28). Water stored in these deposits is then readily available throughout the year. This water will have been partially purified by filtration as it passes through the sand. In the dry season, although the surface of the sand may be exposed, the water below it will be protected from pollution and evaporation.

As shown in Figure 28, a sand dam can be built in stages. It is only raised a small amount each year. This is to ensure that mainly sand and gravel are deposited behind the dam. The finer particles of silt and clay are carried away in the overflowing water rather than forming impermeable layers in the deposits. Since the dam is designed to overflow, it must be built of concrete or masonry; an earth dam would be washed away. Water can be collected from the sand and gravel using an infiltration gallery constructed in the first year's deposits. Alternatively a well can be used. For water supply to communities downstream of the dam a piped gravity flow system may be possible.

Another option for ephemeral (seasonal) streams is to construct a **groundwater dam** in the permeable deposits in the bed of a stream. This prevents the water which is below the bed flowing away. Instead it is stored ready for extraction by an infiltration system or well. An advantage of this type of dam is that it does not project above the bed of the stream and so does not cause surface water flooding.

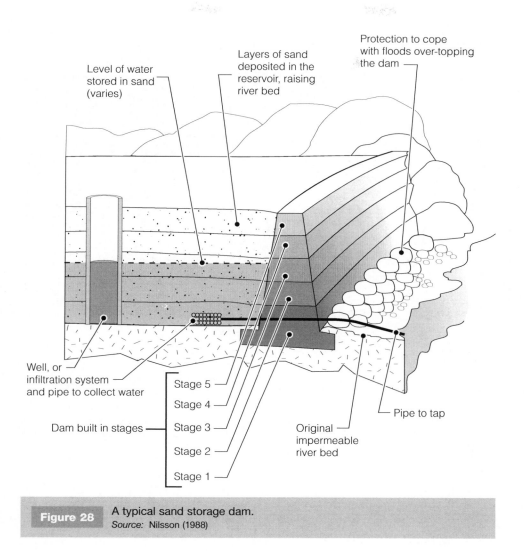

Figure 28 A typical sand storage dam.
Source: Nilsson (1988)

3.4.3 Suction assisted infiltration

Some infiltration systems are connected to pumps (handpumps or motorized pumps) to increase the rate at which water can be extracted from the soil. These systems should be used with care since a fast flow rate often reduces the effectiveness of the filtration process because it draws finer materials deeper into the filter, restricting the flow and increasing flow velocities further. The water from such filters will be of a much better quality than that of the source, but may not be free from disease-causing organisms.

One such system is the SWS filter, which can be buried in the bed of the stream. It is basically an inverted reinforced plastic box with a slotted plate (or 'septum') in it (Figure 29). The slots in the plate are specially made to

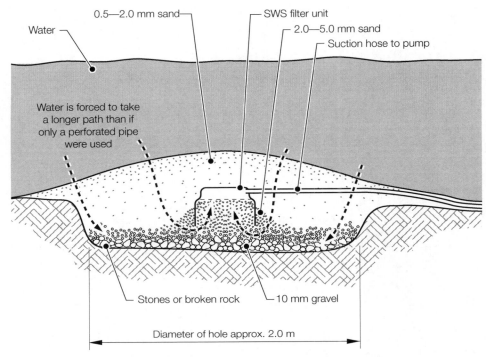

Water

0.5—2.0 mm sand

SWS filter unit

2.0—5.0 mm sand

Suction hose to pump

Water is forced to take
a longer path than if
only a perforated pipe
were used

Stones or broken rock

10 mm gravel

Diameter of hole approx. 2.0 m

This arrangement may be suitable for a pond with an impermeable bed.
The Filter would need to be buried deeper in a stream bed
to avoid damage from floods.

Figure 29 The SWS filter.
Source: SWS (1992)

taper, widening towards the gap above the plate. The taper reduces the risk
of them clogging with particles carried in the water since these either remain
outside a slot or are carried right through it.

Before it is installed the SWS filter is filled with fine gravel and very coarse
sand, and a temporary cover is tied onto it. The box is then inverted on a layer
of granular material placed in a hole excavated in the bed of a surface water
source, and the temporary cover is pulled out from under the box. Then the
hole around the box is backfilled with other suitable granular material to bury
the box and complete the filter. Alternatively, in still water sources (like ponds),
it can be placed on the bed, buried in a mound of suitable granular material
(Figure 29). In both situations, after the box has been buried, heavy pumping
should be used to build up a graded filter around the box. It then works in a
similar way to an infiltration gallery when an ordinary pump is fitted.

If there is no danger of it being washed out during floods this SWS filter can
be used permanently under a stream bed. If it is used under a body of still
water such as a dam or a lake, the bed may become clogged. The best site for
this type of filter is one with a coarse sandy bed.

3.4.4 River and pond intakes

If it is not feasible to take water from the bed of a body of surface water, it can be collected directly by some kind of fixed or floating intake structure. The raw surface water collected from such an intake will usually need treatment, and the design of the intake arrangement needs to be carefully thought about. It is therefore advisable to seek the advice of an engineer. A wrongly designed structure can soon be carried away by floods, clogged by silt, or left high and dry when the water level falls.

It is not a good idea simply to drop a pipe into a river to form an intake. A fixed intake pipe should be held firmly, well above the level of the river bed. This will prevent it becoming buried in deposits. Its intake end should be covered with some kind of grid or mesh to keep out leaves and other rubbish and be accessible for periodic cleaning. Any exposed sections of the pipe should be protected from potential damage by trees or other debris carried downstream. Protection can be provided by concrete, a pile of large rocks (Figure 30), or a heavy wooden frame held by stakes driven into the bed of the stream.

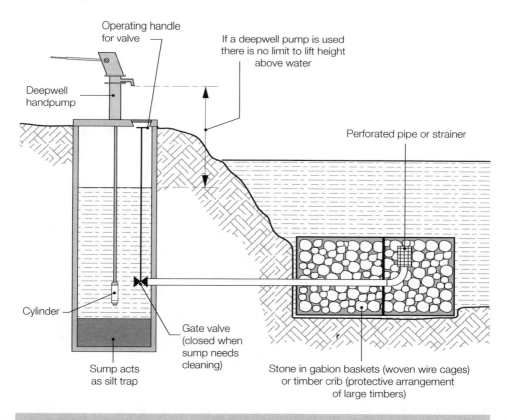

Figure 30   Fixed gravity intake with sump and handpump.
Source: Pickford (1991)

The level of an intake should not be too close to the bottom of a source, because this is usually where there is the highest concentration of suspended solids. Also it should not be so high that it will be exposed when the water level drops because of seasonal changes in flow. A small dam or weir across a stream can be used to ensure that there is sufficient depth of water at the intake throughout the year. Ideally the dam/weir should include a large

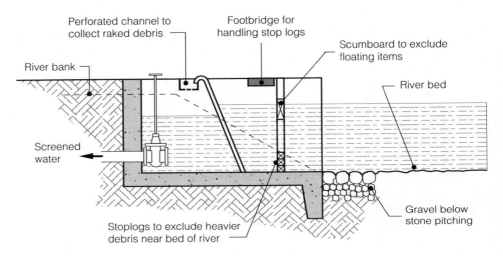

Perforated channel to collect raked debris

Footbridge for handling stop logs

Scumboard to exclude floating items

River bank

River bed

Screened water

Gravel below stone pitching

Stoplogs to exclude heavier debris near bed of river

Section A–A

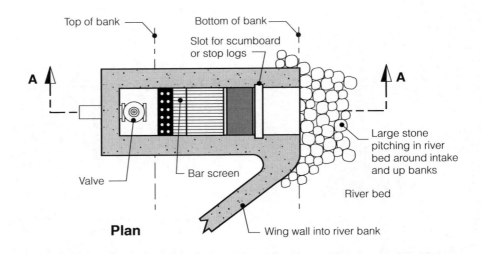

Top of bank

Bottom of bank

Slot for scumboard or stop logs

A

A

Large stone pitching in river bed around intake and up banks

Valve

Bar screen

River bed

Plan

Wing wall into river bank

| Figure 31 | Protected side intake.
Source: Pickford (1991) |

sliding gate or valve positioned at a low level in this weir. Periodically this can be opened to flush though any solids which may be deposited behind it.

A side intake can be built in the bank of a stream. It is best to choose a rocky site where sideways movement of the stream is controlled. It should also be on a straight section, or on the outside of a bend. This is because most deposition that could block an intake occurs on the inside of a bend. However, if the banks on the outside of the bend will be subject to continuing erosion it will not be a good site. An inclined bar screen (Figure 31) can be used at the intake to strain out coarser debris, but trapped debris needs to be raked off the screen at regular intervals so that it does not block it. A high level scumboard can be used to reduce the likelihood of floating materials entering the intake.

A floating intake is particularly suitable for rivers or ponds where the water level will vary substantially. Flexible pipes allow the level of the connection to change with the water level. If the intake hangs from the float about 0.5 m below the surface it should avoid collecting floating debris. It will also collect water containing fewer suspended solids than are found in water deeper below the surface. A coarse screen can be provided around the floating intake as long as it can be periodically cleaned. In rivers the intake will need to be protected from damage during floods.

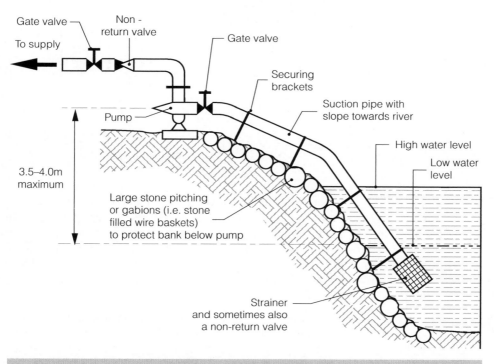

| Figure 32 | Fixed suction intake and powered pump. |
| | *Source:* Pickford (1991) |

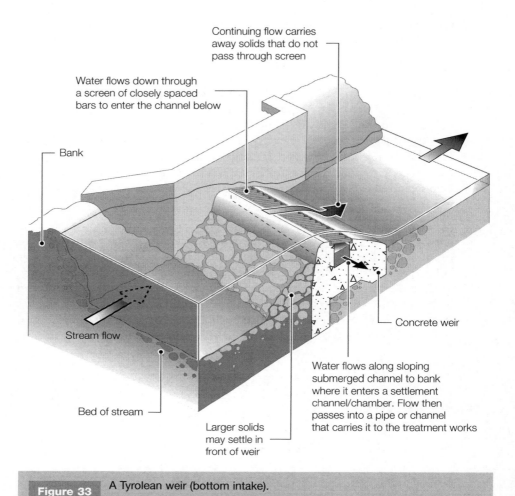

Continuing flow carries away solids that do not pass through screen

Water flows down through a screen of closely spaced bars to enter the channel below

Bank

Stream flow

Bed of stream

Larger solids may settle in front of weir

Concrete weir

Water flows along sloping submerged channel to bank where it enters a settlement channel/chamber. Flow then passes into a pipe or channel that carries it to the treatment works

Figure 33 A Tyrolean weir (bottom intake).

If a well-like reservoir is provided in the bank of a surface water source, water can gravitate from the intake into the well. It can then be extracted from the well by bucket and rope or handpump (Figure 30). There is a limit on the depth from which the water can be lifted by suction. With centrifugal pumps this may be as little as 3.5 m below the pump inlet (Figure 32). More is said about this limit in Section 4.5.2. If the pump is below water then this limit does not apply.

One type of intake suitable for shallow mountain streams which do not carry much suspended solids is the bottom intake or Tyrolean weir. This consists of a sloping perforated plate, or a screen of closely spaced rods, positioned over a channel in the bed of the stream (Figure 33). The size of the holes, or the spacing of the bars, is chosen to limit the size of solids that can enter the channel, which should be freely draining. The channel discharges the

collected water to a suitable point, such as a sedimentation tank (see Section
6.3). The larger particles that become trapped on the Tyrolean screen are
likely to be periodically washed on down the stream, so the screen is to some
extent self-cleaning.

3.5 Disinfection after construction

Before it is used for potable water supply, a well, borehole, spring box, storage tank, pipe system or pump should first be disinfected to kill off any
harmful organisms that are present. Disinfection may also be necessary after
people have carried out maintenance. Section 6.5 describes the use of disinfection for water treatment whereas the following section mainly refers to its
use immediately after construction.

The easiest disinfectant to obtain is usually chlorine, which can be found in
bleaching powder (chlorinated lime), high test hypochlorite (HTH) powder
or tablets, liquid bleach (sodium hypochlorite) and Javel water. The amount
of chlorine each of these compounds contains depends on the manufacturer
and on how well the compounds have been stored. Exposure to air, moisture
and even light will reduce the amount of chlorine available. Therefore these
chemicals should be stored in cool, dry, dark places. Typically, when fresh,
proportionately by mass, bleaching powder contains about 35% available
chlorine, HTH 70%, bleach about 5%, and Javel water about 1%. Even
when properly stored, some of these compounds can lose over half their
strength in a year.

Chlorine reacts with organic matter it comes into contact with, leaving less
chlorine available for killing pathogens. This means that a higher dose will
be needed in water containing organic impurities. The objective is to have
sufficient chlorine in contact with the pathogens for enough time in order to
kill them. In water the free residual value (i.e. final amount of active chlorine
still available) needs to be above 0.3 mg of chlorine per litre (0.3 mg/l,
0.00003% or 0.3 parts per million by mass) after 30 minutes. As mentioned
below, when disinfecting structures a much higher concentration is usually
used.

A useful disinfection liquid is a 0.2% solution of chlorine by mass. To prepare 25 l (about two bucketfuls) of a solution containing 0.2% chlorine by
mass (2000 mg/l) add either 150 g of fresh bleaching powder (ten heaped
tablespoonfuls) or 70 g of fresh HTH to 25 l of water. Alternatively add 1 l
of fresh liquid bleach to 24 l of water, or add 5 l of fresh Javel water to 20 l of
water. Mix the solution well. If using bleaching powder, stir the solution for
10 minutes, allow the solids to settle, and then pour off and use the clear
liquid, leaving at the bottom of the bucket the white sediment which can be
disposed of by burial.

To disinfect a borehole pour in enough 0.2% solution to ensure that, when
diluted by the water already in the borehole, there is sufficient chlorine

present to kill off any pathogens. The volume of 0.2% solution needed can be calculated as shown in Box 12 for a well. After the solution has been added to the borehole operate the pump until the discharged water smells distinctly of chlorine, and then wait for one hour before pumping again. Repeat this procedure of pumping and waiting several times, and then leave the well unused for 12 hours. At the end of that time, pump the water to waste until it does not smell of chlorine.

For a new hand-dug well that is still de-watered you can use a similar procedure to that described on pages 67–68 for the spring box and storage tank, scrubbing the walls between the current water level and the highest level the water is expected to reach. There should be appropriate safety precautions (Section 3.3.4) to allow people at the surface to quickly pull out the worker scrubbing the walls of the well should this be necessary. If the well is full of water you need to use the method described in Box 12.

BOX 12   The amount of chlorine needed to disinfect the water in a newly-constructed well

First you need to find the volume V (in m³) of the well. This is found from the depth of water in the well, H m (which can be measured using a piece of string and a stone), and the diameter of the well, D m.

$$V = 3.14 \times H \times D \times D/4 = 0.785 \times H \times D^2$$

The initial amount of chlorine required to successfully disinfect well water will depend on the amount of bacterial contamination and on the level of other organic contaminants in it. Usually a concentration of 0.01%, (100 g/m³) will be more than enough.

To find the mass of chlorine (M) in grams required to disinfect the volume of water in the well, multiply the concentration required in g/m³ by V in m³.

That is $M = 100 \times V$ grams

To find the volume of the concentrated solution or powder that needs to be used to provide the mass of chlorine needed in the well is best illustrated by an example.

If a well which has an internal diameter of 1.2 m ($D = 1.2$) and a depth of stored water of 2.6 m then

$$V = 3.14 \times 2.6 \times 1.2 \times 1.2/4 = 2.94 \text{ m}^3 \text{ of water}$$

so $M = 100 \times 2.94 = 294$ g of chlorine are needed.

Three example calculations are shown below to show how this amount of chlorine can be obtained from different sources.

- Using the 0.2% solution previously mentioned, which has 2 g of chlorine/ litre, to supply 294 g needs (294/2) = 147 l of the 2% solution. This is about 12 bucketfuls. A smaller number of buckets of a proportionally stronger solution would be easier to prepare and use.

- If bleach with a 5% chlorine content (about 50 g of chlorine/litre) were used, then to obtain 290 g of chlorine, a volume of (290/50) = about 5.8 l of the bleach are needed.

  Although only 5.8 l of bleach are required it would be best to first mix this bleach with a number of buckets of water from the well since this will distribute the chlorine better than if the bleach were added directly.

- If bleaching powder (35% chlorine, or 350 g per 1000 g of powder) were used, then to obtain 290 g the amount needed = (290/0.35) = 828 g of powder. It will be best to mix this powder into a number of buckets of water from the well, and then allow the solids to settle before pouring the clear chlorinated water back into the well.

After the contents of the buckets have been poured into the well, stirring up the water by repeatedly moving a bucket and rope through the water will also help distribute the chlorine throughout the well.

If a well needs to be disinfected because it has become temporarily contaminated then the water level should be reduced as much as possible before disinfection starts. If it cannot be de-watered very much because the work is being carried out during the rainy season and/or the available pumps are inadequate, a lot of water which needs disinfecting may remain in the well. The exposed wall inside the well should be cleaned by scrubbing. After this the amount of chlorine required to disinfect the volume of water remaining in the well should be calculated, as described for a new well in Box 12, and the appropriate amount of 0.2% solution can then be added. If there is a handpump it should be operated until water smelling strongly of chlorine reaches the surface. As with the borehole, the well and pump should ideally be left overnight before water is pumped to waste until the chlorine content reduces to a level which has a taste or smell acceptable to users.

Some chlorine in the well may be effective for a week or more, depending on the size of the well and the pumping rate. However its effectiveness reduces with time.

When disinfecting a spring box or a storage tank it is not normally appropriate to fill them with chlorinated water unless they are very small. Instead, mix up three buckets of the 0.2% solution, and, after draining the tank, scrub the inside walls and base with this liquid.

The person who scrubs the inside of a well, spring box or storage tank should wear goggles to protect his/her eyes from the disinfecting solution. It is also useful to have clean water present to immediately wash eyes that may, despite these precautions, get splashed at some stage. Protective rubber gloves are also advisable. A covered tank should be well ventilated while cleaning work is taking place. After the work is completed any liquid left in the tank should be run to waste through the washout pipe.

3.6 Checking the yield of a source

It is important to know how much water a source can reliably provide. This can be found by carrying out various tests. Ideally these should be carried out at the time of year when the source is likely to yield the lowest amount of water. Measurement at other times will overestimate the amount that is likely to be available. The appropriate time for flow measurement is usually at the end of the dry season. Also remember that over a period of time, if the groundwater in an aquifer is not being sufficiently recharged, the water table may keep on falling year after year. This is a particular problem in regions where irrigation water is being extracted from the aquifer by motorized pumps. It is also a problem in areas which are experiencing less groundwater recharge either as a result of reduced rainfall or of increased run-off because of deforestation.

3.6.1 The safe yield of a well or a borehole

If the source is to be a hand-dug well or borehole, an experienced hydro-geologist who knows the local geology may be able to estimate the expected yield before it is constructed. However, it is only after construction is nearly complete that the actual yield can be estimated based on tests. A good factor of safety should be applied to the results from all yield tests.

Very little water can be stored in a borehole so most of it comes directly from the aquifer. During pumping the groundwater level around a borehole reduces to form a cone of depression (i.e. the groundwater level progressively decreases the closer one gets to the well, roughly similar to the surface of an inverted cone). The rate at which the aquifer yields water depends on the type of soil and the difference in level (the drawdown) between the water level in the aquifer and the water level in the borehole. The maximum safe yield of the borehole is the rate at which water enters the borehole when the drawdown below the seasonally lowest water table stabilizes to an acceptable value.

Unlike a borehole, a hand-dug well can provide considerable storage. This means that for certain periods of the day water can be withdrawn much faster than the rate at which it seeps into the well. The water level will fall during these periods, but not as much as it would in a narrow borehole. For example if people are drawing water during only eight hours of a day they can collect it at an average rate which is three times faster than the average rate with which it enters the well over the whole the day.

It may be unnecessary to check the maximum yield of a borehole or hand-dug well if use of a powered pump is not foreseen. In this case it is only necessary to check that the borehole produces enough water to sustain the handpump. A yield of only 0.2 l/s should be sufficient for most handpumps.

The ability of a well or borehole to provide enough water can be checked by drawing water for eight hours a day for three consecutive days with the pump you will be using. This test is best carried out at the end of the dry season when the water table is lowest. On the second and third day, before starting to pump again, the groundwater level should be checked to see if it has risen back to the same level as at the start of the first day. If the level does not recover, this may indicate that the groundwater source will be depleted. During the pumping test it should not be possible to pump the well or borehole dry. If this test is successful it indicates that the well is unlikely to dry up during normal use. If this did occur it would not only be inconvenient to users but also may cause the pump to wear out very quickly, and break down.

Before deciding when to finish excavating a hand-dug well, it is useful to check the maximum rate at which water is entering it. This is done by de-watering the well, and then stopping the pump and observing the height by which the water rises over a certain fixed period. It is useful to make a number of measurements to find the different rates at which the well recharges for different water depths (e.g. find it for every 0.5 m change in height). From the known diameter of the well and the rise in water level, the volume that has entered the well over any period can then be calculated. If this volume is divided by the time then the average flow rate over that period can be found. The water level will rise fastest when the difference between the water level in the aquifer and the water level in the well is greatest, so the yield from the aquifer will be highest when the well is nearly empty. If this rate is too low to meet the demand at all times of the day, then storage needs to be provided in the well, to supply the additional water needed from water already in it. If the well recharge rates are found for different water levels, then, knowing the volume of water stored per metre depth of well, and the expected pattern of demand, it is possible to calculate a 'water balance' hour by hour throughout a typical day. If this check is made then it is possible to calculate the depth to which the well needs to be dug to ensure that it does not run dry during normal use.

3.6.2 Measuring flow rates

There are various different ways of measuring the rate of flow from water sources or from pumps. For a small unprotected spring the water can be temporarily dammed and discharged from one or more pipes through the dam. The rate of flow from the pipes can then be measured using a bucket and watch. As long as the level of the water behind the dam is stable throughout the test, the sum of the flows measured from each pipe will give the present discharge from the spring. As a result of water lost to seepage under or around the dam, this flow rate might be lower than that which can ultimately be captured.

Approximate discharge rates can also be calculated based on the shape of the discharge from the end of a full flowing horizontal pipe. This method can be used for faster flowing springs, or for measuring the discharge from a pumped system. A formula or tabulated figures are used to calculate the discharge. This information is found in a number of reference books including Technical Brief 27 in Pickford (1991).

For fast flowing springs and small streams a sharp-edged notch can be provided in the top of a temporary dam. This is usually a 90° 'V' notch or a rectangular notch. Reference books including Pickford (1991) give the rate of discharge through the notch based on the depth of water flowing through it.

The flow in larger streams (at least 300 mm deep) can be approximated in m^3/s by measuring the surface velocity (in m/s), multiplying it by the cross-sectional area (in m^2) and a factor of between 0.6 and 0.85. The surface velocity can be established by measuring the time (T) it takes for an almost wholly submerged floating object (e.g. a piece of fruit) to travel a fixed distance (D). The velocity is calculated from $V = D/T$.

4 Raising water

4.1 Introduction

Whatever device is used for raising water, it will have moving parts. These will require regular maintenance and occasional repair. No new water-raising device should therefore be installed unless appropriate arrangements have been made to ensure that this repair work is affordable (ideally to the users) and will be promptly carried out.

A bewildering variety of methods are available for lifting water. Unfortunately, many of them are not suitable for small water supplies because:

- they cannot lift water very high
- they expose the water to the risk of pollution
- they are too expensive to install and operate.

The simplest methods are often the cheapest, and are easier to make and repair using local materials. However, they are sometimes less durable, and usually require more maintenance by the local community. The following sections describe the main methods in order of increasing complexity and cost. Which of these is the most appropriate will depend on the local conditions, the funds available, the feasibility of regular maintenance in the future and the opinions of users.

The first decision to make, though, is whether to use human power for raising water, or to use a motor of some kind. Hand power is suitable for a supply where water is drawn straight from the source, such as a well, and the person drawing water operates the device. If water is to be pumped to a storage tank first, some other type of power will have to be used, such as wind, diesel or electricity.

4.2 Human powered systems of raising water

4.2.1 Hand devices using buckets

The simplest method of raising water is a bucket of some kind on the end of a rope. It is best to use roller bars across the well (Figure 34), a pulley (Figure 35), a shaduf (Figure 36), or a windlass (Figure 37). These devices all ensure that people do not have to lean over the well to raise the bucket, and also make lifting easier. The risk of pollution of the well water will be less if the bucket and rope are never put on the ground. However, with a simple bucket and rope

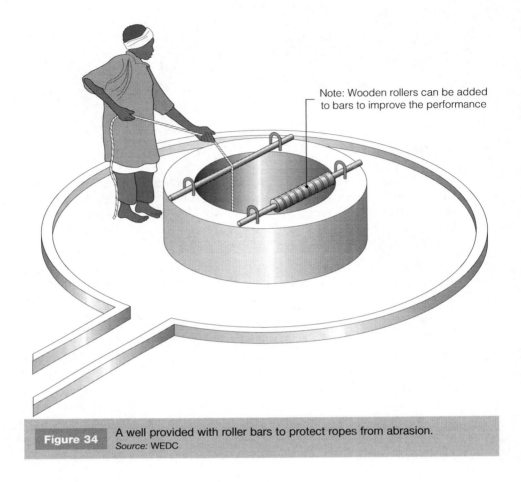

Note: Wooden rollers can be added to bars to improve the performance

Figure 34 A well provided with roller bars to protect ropes from abrasion.
Source: WEDC

system, only users who understand the reduced health risks are likely to adopt this practice. A good hygiene education programme will help increase awareness of these issues. The windlass and the shaduf are good systems because they usually keep the bucket and rope off the ground during use.

The Blair bucket pump (Figures 38 and 39) uses a narrow bucket with a simple valve in its base. This allows it to be used in a borehole. Its simple design means that it can be locally manufactured and repaired, but experience has shown that it is not normally suitable for use by more than 60 people, or where the water is more than about 15 m deep.

4.2.2 Human powered pumps

The devices mentioned in Section 4.2.1 can often be made using materials available in rural areas. Some simple handpumps can also be made by hand in the community from wood, rubber (from vehicle tyres or inner tubes) and plastic pipes. These bucket and rope systems or locally made pumps have the

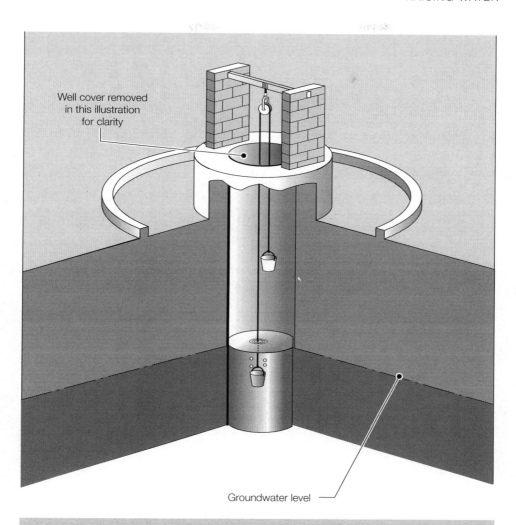

Well cover removed
in this illustration
for clarity

Groundwater level

Figure 35 A protected well with a double bucket system.
Source: WEDC

advantage that their maintenance and repair is well within the capability of the people who made them. However, their disadvantages are:

- they are not usually rugged enough for use by many people
- they can not usually lift water very far
- their design may expose the water they collect to contamination.

In practice a handpump that is manufactured in a factory outside the community, sometimes in another country, will often be used. Such a pump will usually be more rugged and more reliable than a homemade pump, but it will also require regular maintenance and repair. Some of this maintenance requires special tools and a certain amount of skill. For instance, with the

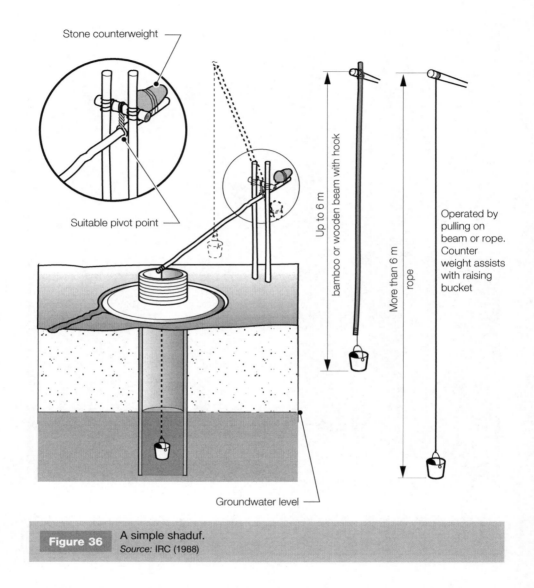

Stone counterweight

Suitable pivot point

Up to 6 m

bamboo or wooden beam with hook

More than 6 m

rope

Operated by
pulling on
beam or rope.
Counter
weight assists
with raising
bucket

Groundwater level

Figure 36   A simple shaduf.
Source: IRC (1988)

traditional design of pump (Figure 44), removing the cylinder from a deep
borehole requires great care, and usually some external assistance. Fortunately, some designs (Figure 45) introduced over the last 20 years now incor-
porate improvements that make maintenance possible without external
assistance. Such types of pump are often termed VLOM pumps, meaning
that they are generally suitable for village level operation and maintenance.
Whatever handpump is chosen, affordable spares must be readily available
locally throughout the life of the pump. This usually needs to be arranged
by national agencies.

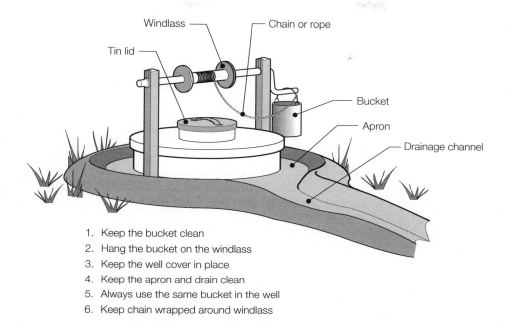

Windlass — — Chain or rope

Tin lid —

Bucket

Apron

Drainage channel

1. Keep the bucket clean
2. Hang the bucket on the windlass
3. Keep the well cover in place
4. Keep the apron and drain clean
5. Always use the same bucket in the well
6. Keep chain wrapped around windlass

Figure 37 A typical windlass and user's instructions.
Source: Morgan (1990)

Reciprocating piston handpumps

Most handpumps are reciprocating piston pumps. These operate on the principle of a valved piston being raised and lowered in a valved cylinder (Figure 40). The piston in such pumps is usually moved by a rod connected to a lever at the pumphead, but a flywheel and crankshaft can also be used to create the reciprocating (up and down) motion. There are three distinct categories of reciprocating piston pump, which are described below. In each category there are many different designs but the names of some well known versions are mentioned for each category.

Suction pumps

These can be of a traditional design (Figure 41) or the more recent rower design (Figure 42). The cylinder to these pumps is usually positioned above ground level and to be regarded as a suction pump the cylinder is always above the water level of the source.

When the piston in the cylinder of a suction pump is moved upwards this creates a partial vacuum below the piston which results in water being sucked up the rising main pipe. In fact the water is pushed into the pipe by the atmospheric pressure acting on the surface of the water source. Because of this reliance on atmospheric pressure these pumps can only lift water from a maximum of about 7 m below the cylinder if the site is at sea level.

Funnel

| Figure 38 | The Blair bucket pump.
Source: Morgan (1990) |

Where the site is at a higher level the maximum lift is less because the atmospheric pressure reduces by 1.1 m per 1000 m change in altitude.

The main disadvantage of these pumps is that if the piston valve seal and piston seal (see Figure 41) are poor, a partial vacuum cannot be formed. The seal provided by the suction valve is also important for the pump to operate properly. In practice water may need to be added to the cylinder to improve the airtightness of these seals before the pump will work. In many situations this 'priming water' is likely to be polluted, leading to contamination of the water which is subsequently pumped through the cylinder. If the pump has a good suction valve it should hold water in the cylinder overnight so that the pump is ready for use in the morning. If the seal is poor then the water will drain out of the cylinder, so priming water will need to be added each

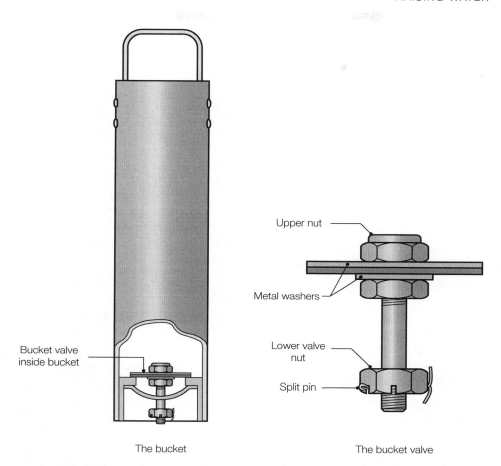

Upper nut

Metal washers

Lower valve nut

Split pin

Bucket valve inside bucket

The bucket   The bucket valve

Figure 39   The Blair bucket and its valve.
Source: Morgan (1990)

morning. If the suction pipe is not airtight the pump will not work at all, or will work ineffectively.

Examples of the traditional suction pump are the Singur and the New No.6. Examples of the rower version are the Rower, and SWS Rower.

Direct action pumps

With these pumps the cylinder is below groundwater level (Figure 43) and the piston lifts the water directly and does not rely on atmospheric pressure. They have no lever handle and the maximum depth from which they can lift water is mainly dependent on the strength of the operator and pump components. It is usually less than 12 m. Good designs allow the piston and the footvalve to be withdrawn through the rising main. Most direct action pumps use air-filled pipes as operating rods but with a few designs water flows inside this pipe rather than around it.

When the rod pulls
the piston **up**:

- V_p closes because of
  weight of water
  above piston

- Water above the
  piston is lifited up
  with the piston

- V_s opens because of
  reduced pressue
  below the moving
  piston

- Water is pumped

When the rod pushes
the piston **down**:

- V_s closes

- The pressure of
  water below the
  piston opens V_p

- Water passes
  through the piston

- No water is pumped
  (unless the diameter
  of the rod is large,
  as in the case of
  many direct action
  pumps)

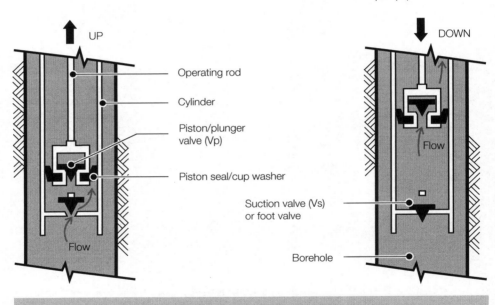

UP

Operating rod

Cylinder

Piston/plunger
valve (Vp)

Piston seal/cup washer

Suction valve (Vs)
or foot valve

Flow

DOWN

Flow

Borehole

Figure 40 How most types of handpump cylinders work.
Source: Pickford (1991)

Examples of direct action pumps include the Tara, the Malda and the Nira
AF85.

Deepwell pumps

Deepwell reciprocating piston pumps can be of a traditional design (Figure
44) and or the more recent open-top-cylinder design (Figure 45). If a deepwell pump is of a strong design it can lift the water from great depth as long
as the operator(s) can apply sufficient force to the handle. However, very
few deepwell handpumps are available to operate beyond about 45 m. An

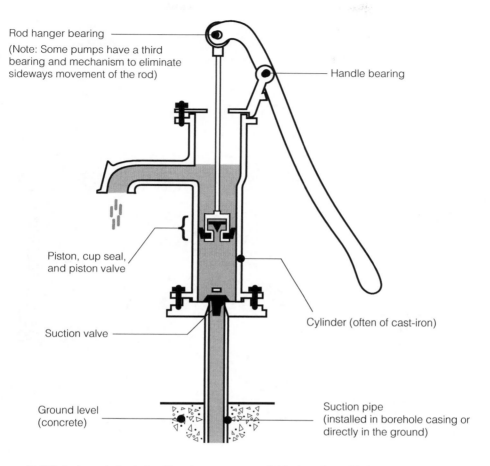

Rod hanger bearing
(Note: Some pumps have a third bearing and mechanism to eliminate sideways movement of the rod)

Handle bearing

Piston, cup seal, and piston valve

Cylinder (often of cast-iron)

Suction valve

Ground level (concrete)

Suction pipe (installed in borehole casing or directly in the ground)

VLOM designs similar to traditional pumps are available, but often with these improvements:

- better suction valves to eliminate priming

- smoother cylinder walls to reduce wear on piston seals

- wear-resistant seal instead of leather (e.g. nitrile rubber) and

- better bearings to prevent the pivot pins wearing out the cast-iron (e.g. using hardened bushes around the pivot pins)

Figure 41 Traditional suction handpump.
Source: Pickford (1991)

open-top-cylinder design makes maintenance much easier, particularly if it is a design which allows the footvalve to also be pulled out through the rising main.

A well known example of the traditional version of a deepwell handpump is the India Mark II pump. Examples of the open-top-cylinder version are the India Mark III and the Afridev pumps.

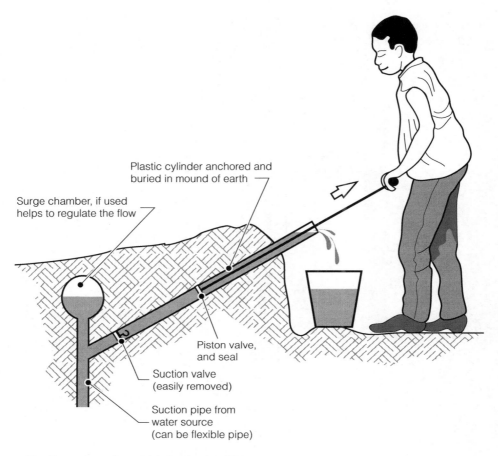

Plastic cylinder anchored and
buried in mound of earth

Surge chamber, if used
helps to regulate the flow

Piston valve,
and seal

Suction valve
(easily removed)

Suction pipe from
water source
(can be flexible pipe)

The Rower pump has the following VLOM features.

● it allows very easy access to the piston and suction valve and footvalve

● it is relatively cheap and easy to manufacture

● on some versions the valves can be replaced using discs cut from vehicle tyre inner tubes

Figure 42 Rower suction handpump.
Source: Pickford (1991)

Force pumps

Some designs of suction and deepwell reciprocating piston pumps can be
used to raise water above the level of the pumphead (e.g. to fill an elevated
tank). Sometimes these are called force pumps. Normal reciprocating piston
handpumps can not achieve this because water leaks out of the pumphead.

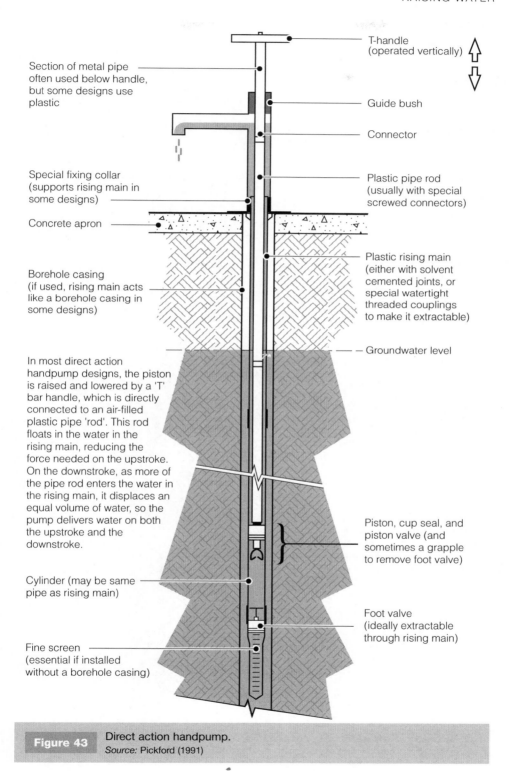

T-handle
(operated vertically)

Section of metal pipe
often used below handle,
but some designs use
plastic

Guide bush

Connector

Special fixing collar
(supports rising main in
some designs)

Plastic pipe rod
(usually with special
screwed connectors)

Concrete apron

Borehole casing
(if used, rising main acts
like a borehole casing in
some designs)

Plastic rising main
(either with solvent
cemented joints, or
special watertight
threaded couplings
to make it extractable)

Groundwater level

In most direct action
handpump designs, the piston
is raised and lowered by a 'T'
bar handle, which is directly
connected to an air-filled
plastic pipe 'rod'. This rod
floats in the water in the
rising main, reducing the
force needed on the upstroke.
On the downstroke, as more of
the pipe rod enters the water in
the rising main, it displaces an
equal volume of water, so the
pump delivers water on both
the upstroke and the
downstroke.

Piston, cup seal, and
piston valve (and
sometimes a grapple
to remove foot valve)

Cylinder (may be same
pipe as rising main)

Foot valve
(ideally extractable
through rising main)

Fine screen
(essential if installed
without a borehole casing)

Figure 43 Direct action handpump.
Source: Pickford (1991)

81

Pump-head:

Most pump-head lever handles work on a similar principle to the handle shown for the traditional suction pump (Fig 41). Some pumps use just one pivot and a chain (or belt) and quadrant system, such as in the India Mk II, shown here.

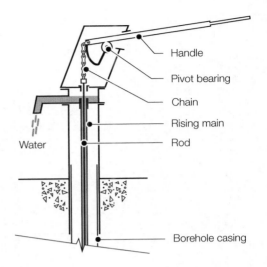

Handle

Pivot bearing

Chain

Rising main

Rod

Water

Borehole casing

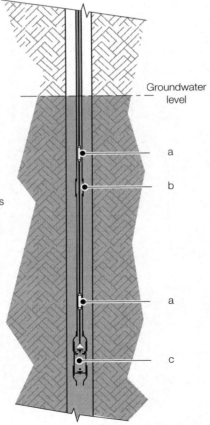

Groundwater level

Rising main and cylinder:

Traditionally, the rising main is of galvanized steel pipes with a smaller diameter than the piston. All pipes and operating rods have to be lifted so that the rod joints (a) and pipe joints (b) can be unscrewed, section by section, to reach the cylinder (c). This operation needs strong people with appropriate lifting and clamping tools, or a mechanized lifting system. Some manufacturers therefore now supply lightweight, thin-walled stainless steel pipes joined with 'rope threads', or plastic pipes with special threaded collars to reduce the weight which needs to be lifted. Rubber 'O' rings can be used to make such joints watertight.

a

b

a

c

| **Figure 44** | Deepwell handpump, traditional design. |
| | *Source:* Pickford (1991) |

Cylinder:
Recent deepwell pump designs have 'open top cylinders' (OTC). These allow the piston (d) to be pulled up through the rising main (e) which is of the same or, preferably, a slightly larger, diameter than that of the cylinder. With these pumps, the piston can be pulled to the surface by pulling out the string of rods.

Rods:
Most rod strings are joined by threaded couplings, but some pumps use special rod joints (f) which can be easily disassembled without tools.

GWL: Groundwater level

Foot valve:
The best designs of OTC allow the foot valve (g) to be removed through the rising main, either with the piston, or by using a fishing tool which is lowered down inside the rising main on a piece of rope after the piston has been removed.

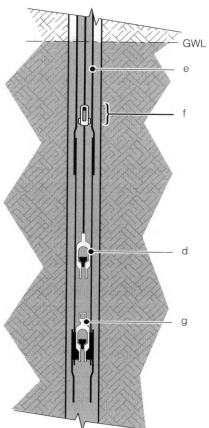

Rising main removal:
In OTC pumps with extractable foot valves, the rising main should never need removing unless the pipe or the lining to the cylinder becomes damaged. Mains with screwed couplings are easily removed.

Should the removal of a solvent-cemented plastic rising main be necessary, sometimes the whole length can be removed by supporting it with tall poles so that it can bend to a large radius curve as it leaves the borehole.

Figure 45 Deepwell handpump, open top cylinder design.
Source: Pickford (1991)

Other types of handpump and footpump

Although most handpumps are of the reciprocating piston type there are some that operate on quite different principles. These include:

- **cylindrical diaphragm pump:** a well known example of this is the Vergnet Hydropump. This is a foot-operated pump that works using hydraulic pressure to inflate a rubber hose positioned inside a cylinder (Figure 46). A hand-operated version, which looks like a direct action pump, and one that uses a converted India Mark II pumphead are now also available.

- **progressive cavity pump:** in this pump the water is lifted by a rotor driven by a rotating rod positioned in the rising main. In the cylinder at the bottom of the rising main, the helical rotor intermeshes with a specially shaped rubber stator to form pockets of water which are lifted from the bottom to the top of the stator and into the rising main. The Monolift pump is one example of this type of pump.

- **oscillating water column pump:** this is a hydraulically operated pump. One unique version, which will now be described, is called the Pulsa pump. It uses a single flexible pipe from a cylinder at the surface to another cylinder below water level. The lower cylinder, which has a valve at its base, contains rubber balls which are hydraulically compressed when the piston in the cylinder at the surface is pushed down. When the piston at the surface is then lifted, the balls expand, pushing water back up the pipe (the polyethylene pipe also contracts to add to this upward movement of water). At the top of its upstroke the piston is lifted out of the cylinder, and the inertia of the water flowing up the pipe causes some water to leave the pipe. At the same time more water is sucked into the bottom cylinder through the valve. The piston is operated by hand and foot via a special lever.

- **rope and washer pump:** in this pump a loop of rope, carrying regularly spaced washers, is continually pulled through a plastic pipe. This type of pump used to only be suitable for wide wells, but in Nicaragua a version suitable for use in boreholes (Figure 47) is now being produced. It can lift water at 8 l/min from 40 m depth. The rope and washer pump has the advantage of a simple design and fairly easy maintenance.

- **Blair bucket pump:** this has already been mentioned in Section 4.2.1.

4.3 Wind powered pumps

The advantage of wind power is that it is free. However, a windmill is required to use it, and they can be expensive. In addition there is the cost of typically about seven days of storage to cope with windless days. Home-made windmills can be made, but they are often not strong enough to last very long under village conditions.

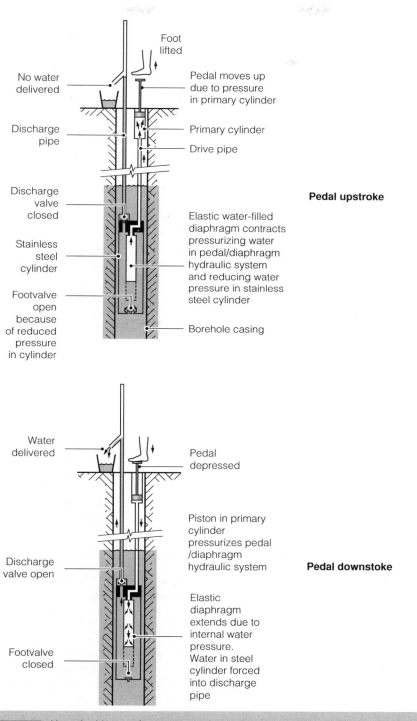

Foot
lifted

No water
delivered

Pedal moves up
due to pressure
in primary cylinder

Discharge
pipe

Primary cylinder

Drive pipe

Discharge
valve
closed

Pedal upstroke

Elastic water-filled
diaphragm contracts
pressurizing water
in pedal/diaphragm
hydraulic system
and reducing water
pressure in stainless
steel cylinder

Stainless
steel
cylinder

Footvalve
open
because
of reduced
pressure
in cylinder

Borehole casing

Water
delivered

Pedal
depressed

Piston in primary
cylinder
pressurizes pedal
/diaphragm
hydraulic system

Pedal downstoke

Discharge
valve open

Elastic
diaphragm
extends due to
internal water
pressure.
Water in steel
cylinder forced
into discharge
pipe

Footvalve
closed

Figure 46   How the Vergnet footpump works.
Source: WEDC

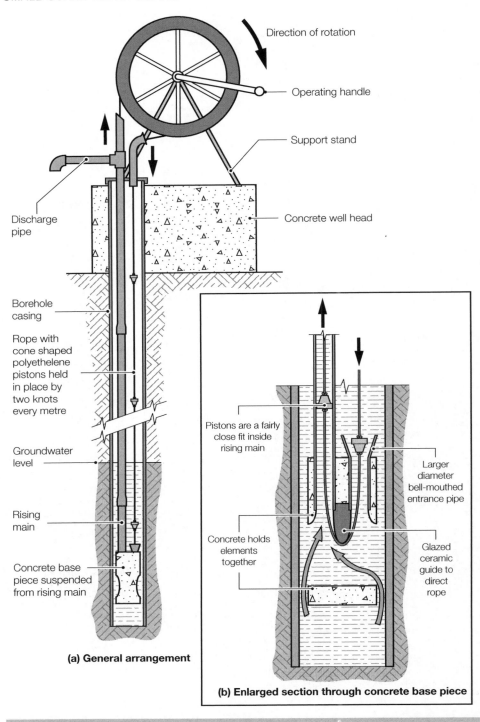

(a) General arrangement

(b) Enlarged section through concrete base piece

Direction of rotation

Operating handle

Support stand

Concrete well head

Discharge pipe

Borehole casing

Rope with cone shaped polyethelene pistons held in place by two knots every metre

Groundwater level

Rising main

Concrete base piece suspended from rising main

Pistons are a fairly close fit inside rising main

Larger diameter bell-mouthed entrance pipe

Concrete holds elements together

Glazed ceramic guide to direct rope

Figure 47	A rope and washer pump used in a borehole. *Source:* WEDC

Before choosing a wind powered pump it is necessary to check whether the wind blows regularly enough and at sufficient velocity to give a reliable supply of pumped water. The average wind speed needs to be at least 2.5 m/s at the level of the rotor. It is important therefore to get expert advice.

Most windmills use a camshaft connected to a rod which drives a piston in a cylinder. This piston acts in an identical way to the piston in a reciprocating piston handpump (Figure 40). In fact a windmill may be combined with a handpump so that water can be pumped by hand if there is no wind. Alternatively, a diesel engine could be provided for when the windmill is not pumping, but this will make the scheme even more costly.

A wind powered pump will only be sustainable if skilled people are available to repair it, and the spares are readily available and affordable.

4.4 Water powered pumps

A hydraulic ram pump (Figure 48) uses the energy of a large volume of water falling a short distance to pump 1–10% of it to a much higher level. Water flows from the source to the pump in a large diameter drive pipe. The pumped water leaves it in a smaller diameter delivery pipe. Both pipes need to be able to cope with high pressure and are usually made of steel.

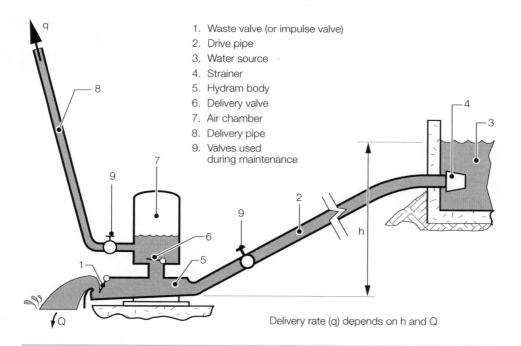

1. Waste valve (or impulse valve)
2. Drive pipe
3. Water source
4. Strainer
5. Hydram body
6. Delivery valve
7. Air chamber
8. Delivery pipe
9. Valves used during maintenance

Delivery rate (q) depends on h and Q

Figure 48	A hydraulic ram pump.

Source: Fraenkel (1997)

A waste valve at the pump allows much of the water from the drive pipe to flow to waste. A hydraulic ram pumps water by utilizing the pressure waves caused by the sudden opening and closing of this valve. The closure of the waste valve stops the flow of water down the drive pipe, creating a sudden increase in pressure which is used to push water through the delivery valve and into the delivery pipe. An air chamber attached to the pump acts as a shock absorber and is an important feature of this pump.

The water which is not pumped but which passes through the waste valve is usually returned to the original stream at a convenient point below the pump. Alternatively it can be collected to power another ram pump at a lower elevation. Although ram pumps can be home-made, these are usually not very durable. In a mountainous area a ram pump is useful for lifting water from a fast flowing spring to a community living at a higher level.

Water current pumps use the movement of flowing water to power a turbine. This is connected to a mechanism that drives a reciprocating piston pump or a rotodynamic pump (see below), either directly or via a gear or pulley system. Such pumps are rare.

4.5 Engine and motor powered pumps

4.5.1 Types of motorized pumps

Where a lot of water needs to be pumped a mechanically powered pump is usually used. Most of these pumps are rotodynamic pumps although other types include reciprocating piston pumps, progressive cavity pumps and diaphragm pumps.

If it is to work efficiently the pump chosen needs to:

- match the desired delivery rate
- lift the water through the required difference in level
- overcome the pumping resistance from the system of pipes it has to pump the water through which will vary with the pumping rate (see Section 7.4.2)
- suit the amount and type of suspended solids present in the water.

Choosing the right pump therefore is a skilled job for which expert advice should be sought.

Some handpumps, particularly those which include rotating elements, such as flywheel operated pumps, or the progressive cavity pump, can be adapted to be driven directly by a belt drive from a motor or engine. This can be a useful stage in upgrading a handpump equipped water supply as the population increases or as it become more able to support motorized systems. However the borehole must be able to yield sufficient water to suit the increased rate of pumping.

Rotodynamic pumps propel water using a spinning impeller or rotor. Centrifugal pumps (Figure 49) are the commonest type, but there are also mixed-flow and axial flow pumps.

Reciprocating piston pumps have cylinders that work in exactly the same way described for the handpumps in Figure 40. The rotary motion of the engine or motor is converted to reciprocating motion using various mechanical devices. They are rarely used for rural water supply.

Progressive cavity pumps have pumping elements that work in the same way as the handpump described on page 84. The pumping element can be mounted vertically or horizontally.

Diaphragm pumps work by using a flexible rubber diaphragm as one wall of a pumping cylinder, which also has an inlet and an outlet valve. A motorized attachment, connected to the diaphragm, moves it in and out to pump the water. Diaphragm pumps are rarely used for water supply. However, they are particularly suitable for dewatering shallow wells or excavations that are less than 7 m deep. Note this type of diaphragm pump uses a completely different operating system to the hydraulically operated system for the Vergnet diaphragm pump mentioned on pages 84–85.

4.5.2 Suction limit and priming

A pump positioned above a water source cannot draw water from more than about 7 m below it, and possibly only from 3–4 m for a centrifugal pump. The actual 'suction limit', as it is called, is very dependent on the atmospheric

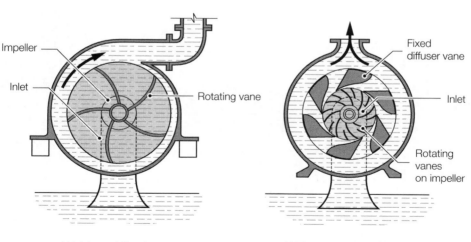

a) Volute centrifugal pump   b) Turbine centrifugal pump

Figure 49	Two types of centrifugal pump.
	Source: Fraenkel (1997)

pressure, reducing with increase in altitude (see comments under suction pumps on pages 75–76).

To start a pump that is positioned above the water level it is necessary to first evacuate the air from it. To achieve this, water is usually added to the pumping element, so that while the pump is moving, air and water are expelled from the pump and the suction pipe until they are both full of water when delivery starts. This process is known as priming. Sufficient water needs to be available to allow evacuation of all of the air in the pipe so it may be necessary to store water for this purpose in a tank which is connected by a branch pipe to the inlet of the pump. A footvalve (non-return valve) on or near the bottom end of the suction pipe (Figure 50) can make priming easier but at this position it can be difficult to access it for repairs. It should be of sufficient size to ensure that flow into the pipe is not appreciably restricted. If the valve is particularly good it should hold water in the pump for some time after it stops, so that re-priming is not necessary. The suction pipe should be airtight or the pump may never start, or will operate very inefficiently.

If the water level will be beyond the suction limit of a surface-mounted pump it may instead be placed below the water level, although the motor may still be at the surface as shown in Figure 51. Provided the pump and its motor are sufficiently powerful, and the delivery pipes are strong enough, then any water that enters the pump can be lifted to almost any height.

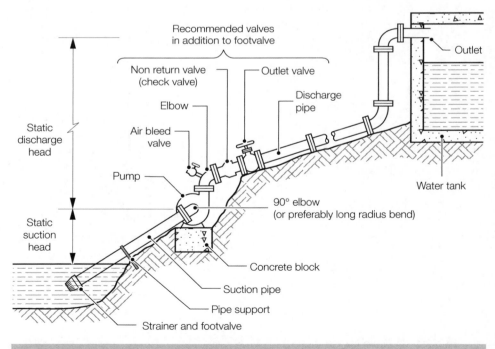

Figure 50	Surface mounted centrifugal pump installation.
	Source: WEDC

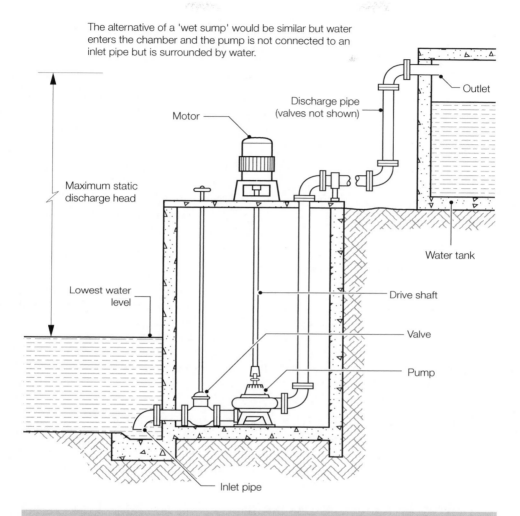

The alternative of a 'wet sump' would be similar but water enters the chamber and the pump is not connected to an inlet pipe but is surrounded by water.

Motor

Discharge pipe
(valves not shown)

Outlet

Maximum static
discharge head

Water tank

Lowest water
level

Drive shaft

Valve

Pump

Inlet pipe

| **Figure 51** | Below surface centrifugal pump installation in a 'dry well'. |
| | *Source:* WEDC |

4.5.3 Power sources for motorized pumps

Pumps can be driven either by internal combustion engines or by electric motors. The internal combustion engines are usually diesel powered although occasionally petrol engines are used. The performance of internal combustion engines is affected by air temperature and altitude. Electric motors can be powered by electricity from the mains supply, from a generator, or from solar energy converted to electricity by photovoltaic panels. Electric motors need less maintenance, are more efficient and are usually more reliable than diesel engines, so they are preferable where a reliable supply of electricity with a stable voltage is available. Generator powered submersible electric pumps capable of handling a high concentration of solids are sometimes used during construction of hand-dug wells that are deeper than the suction limit of surface mounted pumps. For safety reasons these submersible pumps should operate at low voltage (e.g. 50 V).

Matching engines or motors to pumps and arranging efficient ways of transferring power between them is a job best left to an expert. It is therefore best if the engine/motor and pump are supplied together already mounted on a baseplate by the manufacturer.

Recent developments in photovoltaic panels and solar powered pumps mean that this method of pumping water is becoming less expensive, more reliable and more popular, although it is still rarely used. Commonly the panels power submersible centrifugal pumps. These systems have low running costs but maintenance of the electrical components is a highly specialized task.

4.5.4 Sustainability of motorized systems

It is likely to be much more difficult and more expensive to maintain a motorized pump than it is to maintain a handpump so such a system should only be chosen where it can be sustained. In some areas the poor supply of fuel (especially in the rainy season), the poor reliability of electricity (e.g. variable voltage and regular power cuts), or the low level of skills for maintenance and repair will mean that such systems are most unlikely to be appropriate. The water source must also be able to provide the required yield throughout the design life of the system.

In choosing a pump, it is important to select one that a local mechanic employed by the community will understand and for which he will be able to buy spare parts. If there are already some reliable pumps in the area, it is probably best to buy similar ones, made by the same manufacturer. This also applies to the motor or engine that drives the pump.

Internal combustion engines need plenty of maintenance, and can be very expensive for a rural community to run. It is particularly important, therefore, that the community has agreed how money will be raised for operation and maintenance, which includes replacement of air filters, oil filters and fast-wearing parts, regular servicing and repairs. It is also important to decide beforehand exactly who will operate the pump, who will be responsible for repairing it if it breaks down, and from where the fuel or reliable electrical power can be obtained.

These points may appear obvious, but they are all regularly forgotten and explain many of the failures in water supplies throughout the world.

5 Storage

5.1 Introduction

In general there are two types of water supply storage: first, large, unprotected reservoirs such as dams, or smaller rainwater storage tanks, whose main purpose is to store water through dry periods of the year, and second, tanks for storing water ready for consumption during the peak demand periods of each day or during emergencies. The need for storage of rainwater has already been discussed in Section 3.2, and in Section 3.3.2. Box 9 has highlighted the benefits of storing water at low-flowing springs. The use of storage tanks to improve water quality is discussed in Section 6.3.

A reservoir which holds water for potable purposes must be protected from pollution and should be covered to:

- stop windblown contamination entering the tank
- reduce evaporation
- discourage the growth of algae
- to stop insects and animals entering it.

Water should be drawn from the reservoir in a way that avoids contaminating it. A mesh should be provided on the overflow pipe and any ventilators to a covered tank to stop mosquitoes entering and breeding in it. Arrangements should be made to channel any overflowing water away to prevent it causing a boggy area, creating mosquito breeding sites or starting an erosion gully.

Water from an open reservoir should be considered the same as from surface water, and will probably require treatment.

5.2 Dams

A large capacity reservoir can be built cheaply by building an earth dam across a suitable ravine or erosion gully. However, earth dams can be very dangerous if the water overflows in an uncontrolled manner, causing them to be quickly washed away, and releasing huge quantities of water onto the area downstream. The reservoir may also provide a breeding site for mosquitoes (increasing the risk of the transmission of malaria) and or other vectors (such as those that are involved in the transmission of bilharzia and guineaworm disease). Some reservoirs formed by damming streams become filled with silt in only a year or two. Others never fill up because the ground below them is

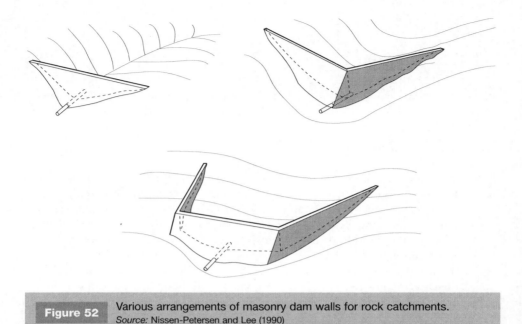

Figure 52 Various arrangements of masonry dam walls for rock catchments.
Source: Nissen-Petersen and Lee (1990)

too porous and/or there are very large losses from evaporation. For these reasons an engineer should be asked to help plan the building of a dam.

Dams on firm foundations can be built from concrete or stone masonry. Where the foundation is not firm a compacted earth dam or rock embankment dam with a waterproof vertical clay core in the centre is more likely to be appropriate. Two or three dams can be used to form the sides of a wedge-shaped reservoir on sloping surfaces which do not have depressions suitable to form a reservoir from a single dam (Figure 52).

Two special types of dam, the sand storage dam and the groundwater dam have already been mentioned on page 58.

5.3 Storage Tanks

5.3.1 General

Reservoirs, in the form of tanks, can be built on the ground surface, raised above it, or below ground level. They can be constructed from various materials. Only three examples are mentioned below in Sections 5.3.2–5.3.4. It is recommended that the appropriate specialist advice is obtained before constructing one.

If a reservoir is built partly buried in firm ground the soil can help to support the sides. It also keeps the water cooler in hot weather. However, one disadvantage is that a pump may be needed to draw the water hygienically from the tank. An alternative is to provide steps in an excavation outside the tank, to give access to a tap near the bottom of the tank. The roof of an underground tank should always be at least 300 mm above ground level, to prevent the danger of surface water running into it.

The outlet pipe should be at least 100 mm above the bottom of the tank so that when the tank empties any silt that has settled out is not all washed into the supply. At the bottom of the tank there should be a drain pipe with a locked valve. This can be opened to completely empty the tank when necessary (e.g. for cleaning). If possible the floor of the tank should slope gently towards the drain. The drain should discharge to a suitable surface water drain or a soakaway (a large stone-filled pit).

All storage tanks require a small amount of maintenance. They should be emptied and cleaned once a year, and leaks repaired. Particularly where a continuous supply is required it is a good idea to use two tanks to provide the volume of storage needed, so one can continue to supply water while the other is being cleaned. Small leaks in a tank above ground may not be serious, but in a buried tank they may lead to pollution entering the tank if the surrounding groundwater level is higher than the water level in the tank. If the water table is high there is also a strong risk that an empty tank will begin to float out of the ground.

A simple approach to determining the volume of storage to be provided is to store one day's supply of water. A more precise estimate of the minimum volume of storage needed for normal supply can be determined by finding the cumulative inflow and outflow for given periods over 24 hours. For example, these figures may be found for each hour of the day. If over the whole day there is just enough water stored to meet demand, the minimum volume of storage required is the largest difference between the cumulative inflow and cumulative outflow figure. This is true provided the exercise commences at a time that ensures for the rest of the day the cumulative inflow is always greater than the cumulative outflow.

Cylindrical tanks are usually more cost-effective and stronger than rectangular tanks.

Storage tanks for potable water should be disinfected after construction, and also after maintenance if there is a risk that they have been contaminated (see Section 3.5).

5.3.2 Brick or masonry tanks

Only an experienced builder should build brick or masonry tanks. The standard of construction and the materials used have to be better than for a normal building, because they must be watertight. Where appropriate, reinforcing bars or wires should be built into the brickwork/stonework to provide additional strength. Alternatively plain wires or mesh can be wrapped tightly around the outside of the tank and later protected with a layer of cement mortar. The quality of bricks and sand used for building tanks needs to be better than that normally used for buildings. Also the cement mortar used for bricklaying and plastering should be rich in cement, and is usually based on a ratio of 1:3 cement:sand. The inside of the tank is usually waterproofed with one or two layers of this cement mortar. It is also a good idea to finally apply a very thin layer of pure cement to the inside face of the tank. This is thrown onto the surface in the form of a thick (porridge-like) mixture of water and cement powder which is spread and smoothed using a steel trowel.

5.3.3 Ferrocement tanks

The use of ferrocement for water storage tanks in developing countries is becoming increasingly popular. It is a particularly simple and cost-effective method of construction. Ferrocement consists of cement mortar reinforced with layers of welded and/or woven wire mesh, sometimes with the addition of plain wire hoop reinforcement for added strength. The mortar mix is usually 1:3 cement:sand by dry volume. The thickness of the reinforced wall of a ferrocement tank is typically less than 50 mm, built up from two or three layers of cement mortar. Fairly coarse clean sand, fresh cement and good workmanship are essential to produce good ferrocement. Curing the fresh mortar by keeping it damp for at least a week is also important to improve watertightness and strength.

If fine wire meshes (<5 mm apertures) are available to cover a skeletal frame made of small diameter reinforcing bars (or welded mesh) then a tank can be constructed without needing formwork (shuttering). With these materials, mortar is pushed into the mesh from both inside and outside the tank. More usually firm shuttering is used on the inside of the tank and the first layer of mortar is applied to the reinforcement from the outside. One interesting method uses external formwork of nylon sacking, held tightly to the outside of the reinforcement by closely spaced spirals of string (Figure 53). With this method the first layer of mortar is applied from the inside of the tank.

Excavations in the shape of a hemisphere, lined with ferrocement and reinforced with chicken mesh and barbed wire (Figure 54) have been successfully used to form very cost-effective water storage tanks, with capacities of 80 m³ or more. Often a low cylindrical wall is built on the edge of the hemisphere to store additional water above ground level and an arched ferrocement roof is constructed across the tank, supported by a concrete-filled pipe in the centre of the tank.

Ferrocement can be added inside any existing leaking tank to waterproof it.

5.3.4 Cement-mortar jars

For storage volumes up to about 2 m³ unreinforced cement-mortar jars can be used, although a spiral of wire is a recommended precaution for the larger jars. Small jars are ideal for domestic storage inside the house. Where large

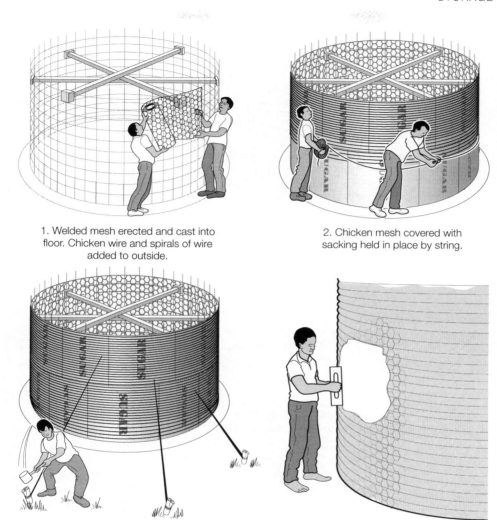

1. Welded mesh erected and cast into floor. Chicken wire and spirals of wire added to outside.

2. Chicken mesh covered with sacking held in place by string.

3. Reinforcement stabilized with guy ropes. Inside plastered

4. Sacking removed. Outside of tank plastered.

Figure 53 Using sacking for temporary formwork for a ferrocement tank.
Source: Hasse (1989)

rainwater storage tanks cannot be afforded, a large jar is at least a starting point for provision of storage. Additional jars can be added subsequently, when the owner is more convinced of the benefits of rainwater storage and can afford them.

Cement-mortar jars can be constructed by plastering a sack stuffed with sawdust or rice husks which has been placed on a mortar base. When the

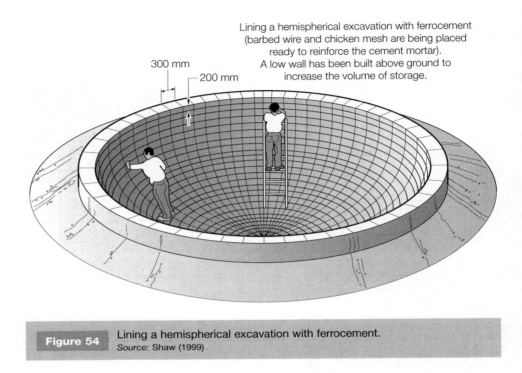

300 mm

200 mm

Lining a hemispherical excavation with ferrocement
(barbed wire and chicken mesh are being placed
ready to reinforce the cement mortar).
A low wall has been built above ground to
increase the volume of storage.

Figure 54   Lining a hemispherical excavation with ferrocement.
Source: Shaw (1999)

mortar layer is hard the sack is emptied and removed. If many jars are being produced then a better method is to use segmented moulds, or shaped blocks to make formwork which can be plastered. The formwork needs to be slightly absorbent for the mortar to stick to it well. Sometimes a layer of mud is applied to the formwork and allowed to dry to give this smooth, slightly absorbent layer. The mud also acts as a de-bonding layer, preventing the cement mortar sticking the formwork. Two layers of mortar are usually used, typically creating a finished wall thickness between 30 and 40 mm.

5.4   Storage in the home

No matter how much care is taken to produce safe water at the source, it will have been useless if it is polluted afterwards. It is therefore very important to protect stored water from contamination. The containers used for storage in a home must be kept clean and regularly rinsed with boiling water or washed out with a solution of one part liquid bleach to five parts water.

Each container should have a cover that fits closely enough to keep out light, insects, dust and other impurities. When possible, it is best to use a container with a small mouth so water has to be poured out rather than drawn by cups or hands dipped into it. Alternatively, draw water from a tap near the bottom of the container. If a container is too large for pouring, and cannot be fitted with a tap, it may be possible to use a siphon tube to draw off the water. Another good idea is to use a long-handled ladle to draw drinking water from the pot. When not in use this ladle can be hung over the edge of the pot, inside it and covered by the lid so it can not become contaminated.

6 Water treatment

6.1 Introduction

Unfortunately, there is no such thing as a simple and reliable water treatment process suitable for small community water supplies. It is always preferable to choose and collect water from a source that provides naturally pure water, and then to protect it from pollution, than to treat contaminated water. Where there are no pure sources, then efforts should be made to reduce the amount of contamination that can reach a source so that subsequent treatment processes have to deal with a reduced amount of pollution. This short book can not give sufficient information for the design of a treatment system. Additional advice should therefore be obtained from one of the sources mentioned in the appendices and references, or from a water engineer.

As mentioned in Section 1.4, treated drinking water must be handled hygienically to minimize the risk of it becoming contaminated again. However, providing a good quantity of water which is not of a very high quality may result in greater health benefits than providing only a small amount of potable quality water. If a community is willing to distinguish between water for different uses, it may be feasible to treat only water which will be drunk or used for food preparation. This is particularly useful when treatment is taking place at household level, but can also apply in other situations such as a community choosing to continue to use a surface water source for laundry and cattle watering, etc.

Table 4 shows the main treatment stages for surface water used in conventional treatment systems. Some of these processes are inappropriate for small rural supplies because of the high level of operational skill required or the need for powered pumps, complicated equipment and chemicals. Treatment should only be considered if it can be afforded and be reliably operated. Both these requirements seriously limit the number of feasible treatment options for small rural communities. Suitable options are discussed below and the treatment of water for individual households is also mentioned where appropriate.

6.2 Screening

Screening of surface water has already been discussed in Section 3.4.4 and of rainwater in Section 3.2.

Table 4 Main treatment stages for surface water

Stage	Name of process	What takes place	Basis for process
1	Screening	Removes large solids (e.g. leaves, pieces of wood).	Physical
2	Aeration	Increases the oxygen content of the water; oxidation of some chemical compounds to an insoluble form; removes some sources of odour and taste.	Chemical
3	Sedimentation	Removes small suspended solids (e.g. sand, silt and insoluble chemicals).	Physical
4	Chemically assisted sedimentation (includes coagulators and flocculators)	Adds suitable chemicals (coagulants) to remove very fine suspended colloidal particles (e.g. clay particles).	Chemical and physical
5	Filtration (various forms, particularly rapid filtration and slow filtration, with one or two stages)	Removes remaining suspended particles; reduces or eliminates bacteria and other pathogens.	Mainly physical. Some biological (particularly for slow filtration)
6	Disinfection	Disinfects the water to kill any remaining bacteria or other pathogenic organisms, and to protect the water before it is consumed.	Chemical

Note:
This table is a simplification of the processes. Large particles are removed first, then smaller and smaller organic and inorganic particles.
Depending on the quality of the raw water, it may not be necessary to use all of the treatment stages shown, or it may be necessary to use some alternative forms of treatment.

At a domestic level straining surface water through a fine cloth will remove the very small water creatures that carry the guineaworm larvae. However the cloth needs to be kept clean, and should always be used the same way round. Sewing a thread onto one side to mark it enables that side to be identified.

6.3 Storage and sedimentation

The simplest method of treatment is storage in a covered tank. If the water can be stored for at least two days, schistosomes (small larvae which cause bilharzia) will die. It will also contain considerably fewer bacteria because these slowly die off because the conditions in the tank are not normally suitable for their survival and multiplication. Pathogens (i.e. disease causing organisms including some types of bacteria) attached to suspended solids will settle to the bottom of the tank together with the solids, further purifying the stored water.

At domestic scale the simple three pot system (Figure 55) can be used to promote settlement during storage.

In a normal storage tank, such as one used to meet peak daily demands, most of the water will not be held in the tank for very long so the degree of improvement in the quality of the water will be limited. If the water level becomes shallow there is always a danger that settled solids will be stirred up by the incoming water and be carried into the distribution system.

Sedimentation tanks can be used to provide conditions suitable for silt and other solid material to settle out of the raw water. Although this removes some of the pathogens it is particularly useful before slow sand filtration to

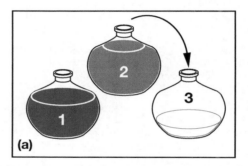

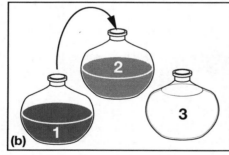 

Drinking water: Always take from pot 3. This water has been stored for at least two days, and the quality has improved. Periodically this pot will be washed out and may be sterilized by scalding with boiling water.

Each day when new water is brought to the house:
(a)   Slowly pour water stored in Pot 2 into Pot 3, wash out Pot 2.
(b)   Slowly pour water stored in Pot 1 into Pot 2, wash out Pot 1.
(c)   Pour water collected from the source (Bucket 4) into Pot 1. You may wish to strain it through a clean cloth.

Using a flexible pipe to siphon water from one pot to another disturbs the sediment less than pouring.

| **Figure 55** | The three-pot treatment system. *Source:* Shaw (1999) |

remove materials that may quickly block such a filter. Sedimentation will not be very effective unless the tank is designed to ensure that the incoming water cannot flow in a narrow stream directly from the inlet to the outlet. Sedimentation tanks therefore often have baffles to spread the flow at inlet and use weirs to collect the flow from a wide area at outlet (Figure 56). If a long rectangular tank is not used, a similar result can be obtained by using a series of vertical walls, to create a long narrow channel which zigzags across the tank from one end to the other. The width to length ratio of a sedimentation tank, or the channel, should be between 1:3 and 1:8. An average horizontal velocity of between 4 and 36 m/hr is necessary to promote settlement, depending on the nature of the sediment and the design of the tank. The

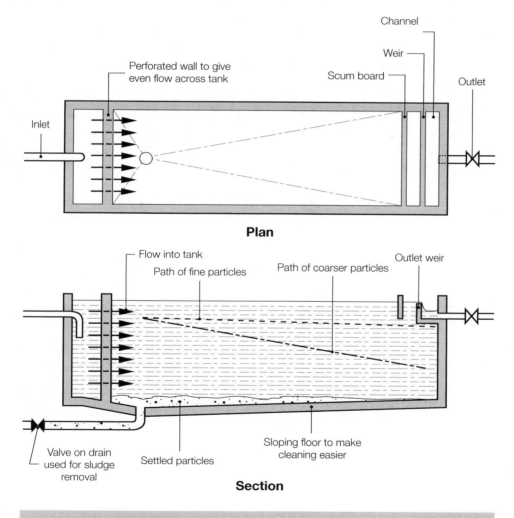

Plan

Section

Figure 56	Plain sedimentation tank.
	Source: WEDC

ratio of the flow rate through the tank to the surface area of the tank usually lies between 0.1 and 1 $m^3/m^2/hr$.

Usually at least two sedimentation tanks will be needed so that one can be in use while the other is being cleaned. Since the tank is designed for silty surface water which has already been exposed to pollution no cover is required unless there is a risk that the tank becomes a breeding place for mosquitoes.

If the suspended solids in the water are very fine, they may not settle quickly enough for simple sedimentation to work. This may be checked by leaving a sample to settle in a bottle for an hour. If the water is still dirty after this time, it is likely to require the addition of special chemicals, called coagulants, to get the fine suspended solids to settle effectively. A frequently used chemical is alum (aluminium sulphate), but a coagulant can also be produced from the seeds of the *Moringa Oleifera* tree that is found in many tropical countries. After coagulation a flocculator may be needed to gently agitate the chemically dosed water before the sedimentation stage.

Large-scale processes of coagulation and flocculation are usually far too complicated to be appropriate for a small rural community and are not covered further in this short book. Alum and ground-up seeds of the *Moringa Oleifera* tree are sometimes used at household level where they are added to buckets of water which are then stirred very slowly for at least five minutes and then left undisturbed for an hour or more for the settlement to take place.

Roughing filters can be a more effective way of removing suspended solids than simple sedimentation tanks (i.e. those that do not use chemicals). The particles removed by a roughing filter are smaller than the spaces between the stones so, despite the name 'filter' the suspended solids are not filtered out. These are rather a type of settlement system in which the water passes through the pores between the stones in a number of chambers. Each chamber is filled with a different size of stones, with the first chamber holding the largest size. The particles that are removed settle on the stones, or are attracted to their surfaces.

There are two main designs of roughing filter, vertical flow and horizontal flow. An upflow system (Figure 57) using three chambers which operate in series is usually considered to be the best type of system. Each needs to be cleaned periodically by quickly opening very large outlets to cause the tank to drain to waste very fast. As a result, most of the deposits are washed off by a high velocity of flow through the pores. In recent years research has taken place to investigate the performance of roughing filters. Helpful guidance on their design and operation is available (see appendices and references).

Roughing filters

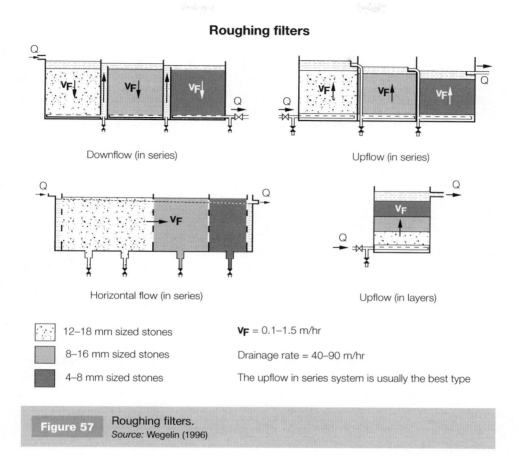

Downflow (in series)   Upflow (in series)

Horizontal flow (in series)   Upflow (in layers)

::	12–18 mm sized stones	V_F = 0.1–1.5 m/hr
▨	8–16 mm sized stones	Drainage rate = 40–90 m/hr
■	4–8 mm sized stones	The upflow in series system is usually the best type

Figure 57	Roughing filters. *Source:* Wegelin (1996)

6.4 Filtration

6.4.1 Introduction

Some kinds of water filter can remove more than 99% of the bacteria and viruses in water if they are correctly operated, as well as other sources of disease such as cysts, ova and schistosomes. Loose granular materials such as sand are often used in these filters. They can treat water on a large or small scale.

At a household scale, instead of using granular materials, a solid porous filter (e.g. ceramic 'candle') can be used to filter the small amounts of water needed for drinking. These filters are rarely used on a wide scale in low-income areas. The ceramic material needs to be regularly scrubbed clean and disinfected using boiling water. It should be replaced if any crack develops or when, as a result of the frequent scrubbing, it wears too thin to filter the water effectively. Some manufacturers include silver compounds in the ceramic which kill micro-organisms that come into contact with them.

In granular filters the medium is usually sand, but other materials such as burnt rice husks may be used. Charcoal is sometimes used to remove tastes and odours but it can become a breeding ground for bacteria and is not recommended for small treatment systems. Whatever granular material is used, the particles should be clean and fairly uniform in size. Different types of filter need different sizes of sand. Rapid filtration operates at a fairly high filtration rate (e.g. 5–7 $m^3/m^2/hr$) and needs fairly coarse sand (e.g. 0.7–1.0 mm particle sizes). Slow filtration, which can remove virtually all bacteria, uses a slower filtration rate (e.g. 0.1–0.2 $m^3/m^2/hr$) and uses much finer sand (e.g. 0.15–0.35 mm). Both use the depth of water above a bed of sand to push water through it into a collection system that covers the whole floor of the tank.

6.4.2 Rapid sand filters

Most rapid sand filters are contained in an open reinforced concrete tank. Usually the depth of water above the sand provides the pressure to push the water through the sand. A pressure filter is a special type of rapid sand filter in which the sand is instead held inside a closed prefabricated, pressure-resistant unit. This means that a pump (or sometimes gravity flow) can be used to increase the pressure of the water contained above sand, resulting in a faster rate of filtration.

Both types of rapid filter can remove most of the suspended solids from water, particularly if coagulation, flocculation and sedimentation precede them, but many bacteria and viruses will still remain in the water. This means that to produce potable water they must be followed by disinfection, or by a slow sand filter.

Rapid sand filters usually need a pumps, and ideally an air compressor too, to regularly backwash and fluidize the sand bed, to wash out the material removed from the raw water. Hence such filters are generally too complicated for operation in small communities. They will therefore not be considered further.

6.4.3 Slow sand filters

Slow sand filters are simpler to operate and maintain than rapid sand filters and are usually very effective in removing bacteria and other water-borne pathogens. Although operation requires little skill, a slow sand filter needs regular attention from a committed caretaker who understands the process if it is to perform well. The filter also needs to be carefully designed. Advice on this can be obtained from some of the sources mentioned in the appendices and references.

A slow sand filter consists basically of a large tank containing a bed of fine sand which is initially 0.9–1.2 m thick. Water flows through this bed to reach a set of drains that take it to an outlet weir (Figure 58). The filter works by a combination of biological action, adsorption (like a roughing filter) and straining. Its most important feature is the sticky deposit (called the *schmutzdecke*) which forms on the very top of the sand. In this layer bacteria and microscopic plants multiply to form a very fine straining mat in the top-most few millimetres. Useful micro-organisms in the mat and deeper in the sand feed on any pathogens in the water. This greatly improves its quality.

If the water going into the filter is reasonably clear, a slow sand filter may run for weeks or even months without cleaning. If the water is not reasonably clear the slow sand filter will need very frequent cleaning. If the water is very dirty, it is advisable to try to improve it before it enters the filter. This is best done using a sedimentation tank or a roughing filter. Where appropriate a rapid sand filter can be used.

To produce good quality treated water the flow rate through the filter needs to be carefully controlled. Control can be by manually adjusted valves at inlet (Figure 58) or at outlet (not shown). A 'V' notch (Section 3.6.2) on the inlet or outlet weir, used together with a depth of flow indicator, allows

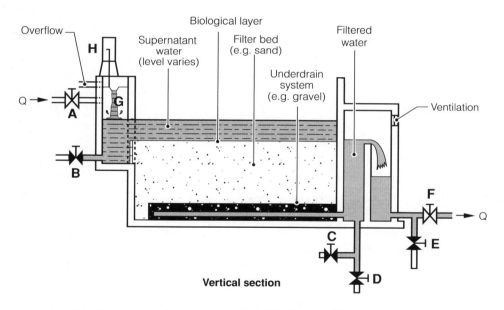

Vertical section

A Valve for raw-water inlet and regulation of filtration rate (discharges behind weir)
B Valve for drainage of supernatant water layer
C Valve for back-filling the filter bed with clean water
D Valve for drainage of filter bed and outlet chamber
E Valve for delivery of treated water to waste when filter is maturing
F Valve for delivery of treated water to the clear-water reservoir
G Inlet weir
H Calibrated flow indicator

Figure 58   Basic components of an inlet controlled slow sand filter.
Source: IRC (1987)

the operator to check the flow rate. Automatic systems can also be used. However, they need to be simple and reliable.

A slow sand filter requires cleaning before the flow rate becomes insufficient for the community's needs, even when the water level above the filter has risen to its maximum depth (e.g. 1.5 m). Cleaning comprises draining the water to about 50 mm below the top of the sand and then scraping off the top 10–20 mm. This material is discarded, or is washed clean and stockpiled for reuse.

When, after successive cleaning, the sand bed is only 600 mm thick, 300 mm or more of clean sand has to be added to the bed, bringing the bed back to its original thickness. First the surface must be scraped clean in the usual way. It is then best to remove a 400 mm layer of the remaining sand, or to move this to one side of the tank, to enable the new sand to be placed low in the bed. The removed sand is then replaced to form the top layer. This is advantageous because useful organisms in the existing sand can then quickly regenerate the *schmutzdecke*.

The quality of the water flowing from a well-operated sand filter is usually very good. However, after cleaning, the filtered water will not be of potable quality until a few days later, during which time the *schmutzdecke* is being re-formed. During this 'ripening' period water should be recycled or run to waste, although it could be used for drinking water if it were disinfected (see Section 6.5). If only one filter is being used, and disinfection is not feasible, then several days of storage of treated water would be necessary. A better idea is to use more than one filter, so the other filter(s) can provide the full daily demand for treated water during cleaning and for a few days afterwards.

Water that is to pass through a slow sand filter should never be chlorinated because the residual chlorine will kill off the useful micro-organisms living in the sand (Section 6.5.2).

Building and operating a slow sand filter properly is not straightforward so before choosing to use one it is best to obtain advice from a water engineer. Useful guidance is available in some of the material mentioned in the appendices and references.

Small sand filters for use by individual families, often based on partly filling oil drums with sand, have been used in some places. These are not able to produce as good a quality of water as conventional slow sand filters. This is partly because of three main reasons:

- the depth of sand is often less than 600 mm
- there is often no flow rate control system
- the maximum depth of water above the sand is usually not very deep (e.g. only 0.3 m instead of 1.5 m) leading to short periods of operation before cleaning is necessary. This period can be extended by using sedimentation, such as provided by the three-pot system (Figure 55), before filtration.

Despite the fact that not all pathogens may be removed, the water quality from a household filter will be much better than the raw water. Hence they are still worthwhile, provided the owner is committed to following the correct operating procedure involved in regularly scraping off the top layer of sand and periodically replacing the sand which has been removed. As mentioned above, proper provision also needs to be made for supplying potable water when the filter is being cleaned and for several days afterwards.

6.5 Disinfection

6.5.1 Introduction

Disinfection reduces the numbers of organisms in water to such a low level that no infection or disease results when the water is drunk. It is a term usually used for the addition of chemicals to water, but ultraviolet (UV) light (which can be from the sun or an electrically powered lamp) also kills some harmful micro-organisms, as does boiling water.

The need to disinfect water to kill all pathogens in drinking water should be considered carefully. Despite consuming clean water users can still ingest similar pathogens, sometimes at much greater concentrations, from other sources, such as with food. Where this is the case the health of the community may improve more if, instead of implementing chemical disinfection, the money is spent on funding a programme which leads to changes in hygiene practices (See Section 1.4).

Disinfection on a regular basis is rarely practicable in rural areas. It should be viewed as a last resort. As mentioned earlier, it is far better to find, protect and use an unpolluted source.

6.5.2 Chemical disinfection

Water supplies are usually disinfected by adding chlorine, although other substances like ozone gas may also be used. Other chemicals, such as iodine and potassium permanganate, may also be used for small-scale disinfection. Sources of chlorine and ways of preparing solutions with certain concentrations of chlorine have already been discussed in Section 3.5.

Chlorine can kill bacteria, schistosomes, some viruses and, in higher doses (>2 mg/l), amoebic cysts. There is little danger to health from excessive dosing, but if too much chlorine is added, the unpleasant taste may drive people to use more heavily polluted water instead.

Chlorination is often an unreliable process when used in small communities. This is because of one or more of the following problems.

■ A reliable source of chlorine may not be available in outlying districts. (However, a recent development is the availability of simple electrically powered devices that can produce dilute sodium hypochlorite from the electrolysis of a solution of common salt. Such devices make regional supply of an easily applied source of chlorine more feasible.)

- The strength of the chlorine compound can vary with age and storage conditions. It needs to be frequently measured.
- The rate at which the compound is applied needs to be reliably regulated to supply the correct dose. Sometimes this rate will need to vary because of changes in the flow rate.
- Chlorine needs to be in contact with the water for sufficient time (the contact time), before the water will be safe to drink. This period is typically 30 minutes.
- The amount of chlorine required for effective disinfection will vary with the quality of the raw water. However, the quality of surface water will depend on the recent pattern of rainfall.
- A reliable method of testing the free residual chlorine is needed. Free residual chlorine is the amount of chlorine still available after disinfection, to kill any pathogens that may subsequently come in contact with the treated water.

Any organic contaminants in the water quickly absorb chlorine; it is therefore important to check that sufficient free residual chlorine is present after the required contact period. A residual amount of 0.3 mg/l (0.3 parts per million) at the point at which the water is collected is usually considered sufficient to indicate that the water has been successfully disinfected. To achieve this residual value of 0.3 mg/l a much higher concentration (e.g. 3 mg/l) is initially needed. Simple kits for measuring chlorine are available. They require a regular supply of chemical tablets.

Chlorination is sometimes applied after slow sand filtration, not necessarily because of remaining pathogens, but to create free residual chlorine in the water to protect it from minor contamination while it is in a piped distribution system. Measuring the free residual value in the water collected at ends of the distribution system can then be used to monitor the state of the pipework. One risk of contamination is from polluted water that may seep into leaking joints or at illegal connections. This can occur if the water pressure in the pipe becomes lower than the groundwater pressure outside the pipe.

Chlorine should never be applied before slow sand filtration because the residual chlorine in the water is likely to kill the useful micro-organisms on and in the sand bed. Dirty or cloudy water should not be chlorinated either, because the dirt in the water will absorb the chlorine. Sedimentation and/or filtration before chlorination will prevent chlorine being wasted. Note that there is very little value in chlorinating if insufficient chlorine is added to produce the required free residual amount.

It is wise to seek expert advice about chlorination. Regular attention and careful adjustment is necessary to ensure systems run reliably. **There is no point in using a chlorinator that is not reliable**. Simple chlorinators, which dispense a chlorine solution at a constant rate, can be bought or made with materials available in most developing countries. One simple system that is rarely mentioned in the literature is the constant head aspirator shown in Figure 59.

Some practitioners use a perforated pot, filled with a mixture of bleach and coarse sand to apply chlorine to wells. The pot is meant to slowly release chlorine into the well. There is considerable doubt about the reliability of such systems, even if the contents of the pot are replaced every two weeks.

Water sterilising tablets are sometimes available for household-scale disinfection but are rarely affordable for regular use in rural areas. A cheaper alternative is to use Javel water, or to use a prepared 1% solution of chlorine in water (See Section 3.5) but correct preparation and/or dosing by householders is likely to be problematic. Three drops of 1% solution should be

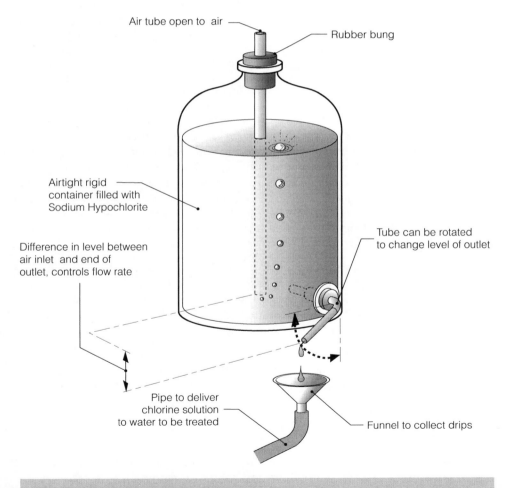

Air tube open to air

Rubber bung

Airtight rigid container filled with Sodium Hypochlorite

Difference in level between air inlet and end of outlet, controls flow rate

Tube can be rotated to change level of outlet

Pipe to deliver chlorine solution to water to be treated

Funnel to collect drips

Figure 59 Constant head aspirator.
Source: WEDC

mixed thoroughly into to each litre of clean water to be treated and the water should then be allowed to stand for 30 minutes or longer before it is drunk. If the strength of the chlorine compound is unknown, add enough of it until it can still be tasted in the treated water after the contact period.

Iodine works in a similar way to chlorine. It can be bought as a tincture (solution) about 2% strong from chemists, and should be added at a rate of two drops per litre and left for 30 minutes or longer before using the water. It is suitable for occasional use, but should not be used continuously for a long time lest it cause any unpleasant side effects.

6.5.3 Solar disinfection

If clear water is exposed to strong sunlight for a sufficient period of time the light kills any bacterial pathogens. This is because the UV radiation in sunlight will destroy most faecal bacteria. The water can be contained in clear glass, or plastic bottles or even plastic bags can be used. Increasing the oxygen content in the water (e.g. by shaking the partly filled bottle before and during exposure) has been found to speed up the die-off rate of the bacteria.

The effectiveness of the disinfection process increases with temperature, although it does not have to rise above 50°C to be effective. To ensure a beneficial temperature rise the SODIS system uses half-blackened bottles that are laid clear side up, typically on a sloping roof. Plastic bottles made of polythylene terephthatate (PET) are recommended but durable plastic bags can also be used. In tropical regions, it has been found that a safe exposure period is about five hours, centred about midday under bright or 50% cloudy sky. This period needs to be extended to two consecutive days under 100% cloudy sky. More information about this method can be found on the SODIS web site, which can be accessed through the GARNET web site mentioned in Appendix 3.

As people are unlikely to want to drink the warm water it should be hygienically stored after treatment to allow it to cool. It is preferable to store it in the same container until it is used, since this avoids the risk of it becoming contaminated again. One advantage of this method of treatment is that, unlike boiling, it makes little difference to the taste of the water.

Some solar treatment systems do not use the UV light but only a temperature rise to kill the pathogens. These systems are often called solar water pasteurizers. Strictly speaking, for pasteurization to take place the temperature should reach 79.4 °C.

6.5.4 Boiling

Boiling of water is often suggested as a method of disinfection of limited amounts of drinking water. A typical recommendation for disinfecting water by boiling is to bring the water to a rolling boil for at least five minutes. It is now thought that boiling for such a long period is unnecessary. Reaching a temperature of 100 °C for a few moments is more than sufficient to kill virtually all pathogens and most are killed before the temperature reaches 70°C.

The main disadvantage of boiling water is the high cost of the additional fuel required. Already many rural areas are becoming deforested and increased use of firewood for boiling water will contribute to the degradation of the environment. Boiling water adversely affects its taste but after cooling the taste can be improved by vigorously stirring the water, or shaking it in a bottle to aerate it. Some users like to add a small amount of salt.

Like the water disinfected by solar methods, boiled water needs to be stored hygienically, while it cools and until it is used.

6.6 Aeration and removal of iron, manganese, tastes and odours

In a few areas, high concentrations of iron and manganese in the groundwater can give it an unpleasant taste, and a brownish colour to clothes washed in it or to rice cooked in it. While they are not harmful, these chemicals may also give the water an unpleasant taste that may discourage people from using it. The concentration of iron and manganese and some other unpleasant tastes and odours can sometimes be reduced by aeration. Aeration usually changes the iron and manganese so that they are no longer soluble in water, and they form sediment that is easily removed by storage or filtration.

On a community scale, aeration can usually be achieved by allowing the water to trickle through a well-drained layer of gravel in a perforated and ventilated container, but it is best to ask a water engineer for advice. On a domestic scale, aeration can be achieved by vigorously shaking water in a partly full container such as a jerrycan. It can then be stored to allow sedimentation to take place.

Iron and manganese removal plants which use aeration followed by filtration are available for attachment to handpumps, but communities rarely maintain them properly and they fall into disuse. Iron can also be removed organically in a slow sand filter.

Oxygen is needed to sustain the useful micro-organisms in slow sand filters. Surface water usually contains sufficient oxygen but groundwater is likely to need aeration before it is filtered.

6.7 Removal of salt

Salty water can be purified by various methods. Small solar stills that are based on the evaporation and condensation of water can be suitable for household use. Other methods of desalination are needed for the volumes of water required by a community but simple methods are not available. The methods used for desalination in developed countries are too complex and costly for use in developing countries.

When ground water is salty as in some flat areas near the sea, there is sometimes fresh water lower down. If so, sometimes a deep tube well or borehole may be sunk to reach the fresh water below.

6.8 Removal of fluoride

Where fluoride is found in concentrations over 4 parts per million, those who drink it risk long term damage to their teeth and bones. The local medical authorities usually can advise if the ground water in an area contains a dangerous amount of fluoride. Fluoride can be removed by the addition of lime and alum, followed by sedimentation. This is known as the Nalgonda technique. Other methods pass the water through granular activated alumina or through bone char. All these methods need specialist advice and long-term support to ensure sustainability.

6.9 Removal of arsenic

Harmful concentrations of arsenic have recently been identified as a major problem in groundwater in Bangladesh, Nepal, Vietnam and West Bengal in India. High concentrations are also found in some other parts of the world. There is no simple test available to accurately measure low, yet dangerous, concentrations of arsenic in water. Work is presently taking place to find such a method. Investigation is also focused on simple ways to reduce the concentration of arsenic in water to a safe level.

Most of the existing methods of removal are hard to sustain without a ready supply of chemicals such as chlorine and alum. One promising method is similar to the SODIS method of disinfection. This SORAS method also uses bottles that are exposed to sunlight. Exposure is followed by a period for settlement of newly formed arsenic compounds. The water above the sediments has a much reduced arsenic content.

Fortunately iron removal also removes much of the arsenic, which precipitates out combined with the insoluble iron compounds after aeration. Some methods therefore add iron compounds to the water as part of the treatment process.

Where arsenic removal is not feasible the water can still be used for purposes other than for cooking and drinking.

7 Piped water distribution

7.1 Introduction

It is not possible in this short book to offer much guidance about the choice of appropriate pipe materials, pipe diameters and pipework layouts for the supply of water. Help can be found in some of the sources mentioned in the appendices and references. A better option is to contact a water engineer for advice.

7.2 House and yard connections

As mentioned in Chapter 2, there are great advantages in piping water into individual households, rather than providing it for collection from widely spaced public water points (standposts). This is because the convenience of having water at the home usually leads to increased usage, which results in an improved level of hygiene and better health. In fact in some situations it may be essential to provide water at homes (either to a yard tap just outside the house, or to one or more internal taps) to obtain any health benefits from the water supply. However, water should not be provided at homes without providing an appropriate way of disposing of the used water too. If the used water is not disposed of properly it can create a severe nuisance and health risks such as increased breeding of malaria-carrying mosquitoes.

House connections are of course more expensive than scattered standposts. However, there is often a high demand for this level of service, which means that people are willing to pay more for it, although it may be necessary to collect the connection charge in instalments. Payment by consumption, registered on domestic water meters, can reduce wastage of water but the use of meters may introduce unwarranted additional costs and increase the complexity of managing the scheme to a point where it becomes unsustainable.

Even if house connections are not provided initially, it will be prudent to design the main pipes to allow houses to be connected to them during the design life of the system. However, this will only be appropriate where there is a source that will be able to meet the increased future demand.

7.3 Public water points

If water is to be collected from public standposts, the taps and the supporting structure should be durable. The design needs to make water collection as easy as possible and the users should be involved in the design process. Enough taps should be provided to avoid congestion at peak times (see Chapter 2).

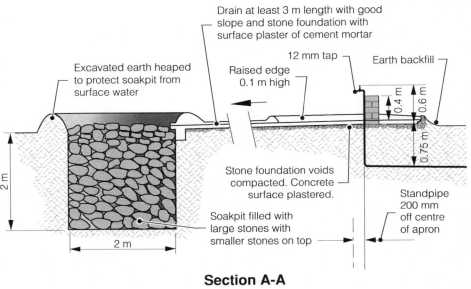

Section A-A

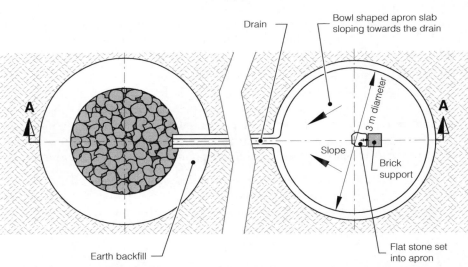

| Figure 60 | An example of a stone-filled soakpit used for a public standpost. *Source:* DLVW (1983) |

Taps are the most frequent component of a piped water system to break down. They should therefore be the strongest type available. Someone whose family collects water from the standpost , and who is willing to look after it, should be issued with spare washers and spanners and be shown how to replace old washers and repair leaks. Self-closing taps, designed to prevent wastage, are rarely appropriate. They often do not last very long when used for public water points, as frustration with them leads to mistreatment and vandalism. It is better to help the community understand the importance of turning the normal taps off after collecting their water so that simpler taps, which are easier to obtain and repair, can be used.

If there is plenty of water flowing from a spring it may be possible to avoid using taps. This will be acceptable only if the waste water which is freely discharged from the pipe provided at each standpost can be disposed of in an appropriate way, preferably to a nearby water course.

If a tap is used then disposal of waste water is still important, although, because the amount of water is less than that from a free flowing pipe, it may be possible to dispose of it into soakage systems. One example of a soakage system is a stone-filled soakaway (Figure 60). Instead of filling the soakage hole with stones it can be left empty and be lined with permeable lining (Figure 61) but such a pit needs to be covered. Other alternatives for disposal of wasted water are shallow soakage trenches (Figure 62) or using the water

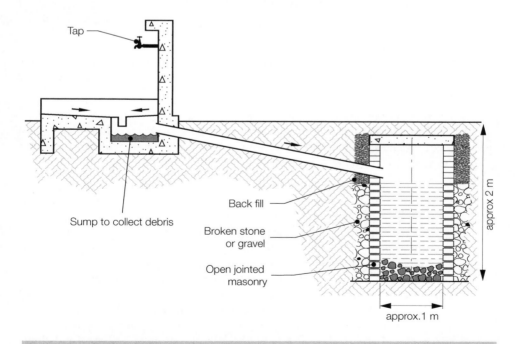

Tap

Sump to collect debris

Back fill

Broken stone or gravel

Open jointed masonry

approx 2 m

approx.1 m

Figure 61	Lined soakaway.
	Source: IRC (1979)

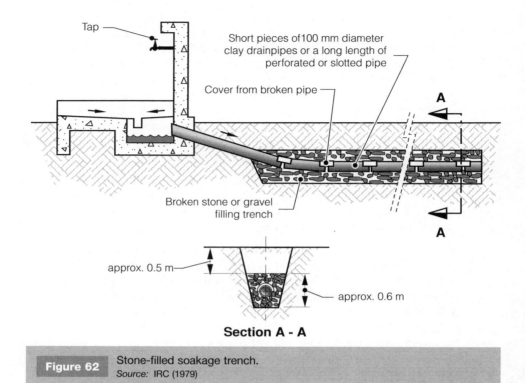

Tap

Short pieces of100 mm diameter clay drainpipes or a long length of perforated or slotted pipe

Cover from broken pipe

A

Broken stone or gravel filling trench

A

A

approx. 0.5 m

approx. 0.6 m

Section A - A

Figure 62	Stone-filled soakage trench. *Source:* IRC (1979)

to irrigate a garden. If waste water is not disposed of properly, muddy conditions around the standpost will deter users and give an opportunity for mosquitoes to breed.

To save people having to carry some water home it may be appropriate to make provision for clothes washing near to the standpost, as long as the used water can be disposed of safely. Some communities may like facilities near the standpost for people to bathe themselves and/or children.

The design of a water point should allow for local methods of carrying water. The taps should be high enough above the platform for containers to fit conveniently beneath them. Where jars with round bottoms are used, a depression might be needed in the platform to give appropriate support. If the water container is customarily carried on the head, it will be convenient to provide a filling platform about 1.4 m above the ground. This avoids some unnecessary lifting after the container is filled. A second, lower platform about 0.45 m from the ground should also be provided for children and old people to use. Where only a low-level tap is provided, an intermediate-level plinth can be built so a user can put their filled container on this before he/she lifts it onto his/her head. Another idea is to use a very high level discharge point, with a short hose attached to it, so a user can fill a container while it is still on his/her head.

7.4 Pipeline materials, design and laying

7.4.1 Pipe materials

To protect its quality, water for potable purposes should flow in pipes and not in open channels, especially once it has been treated. The pipe joints should be watertight, not only to avoid wasting water, but also to avoid the risk of polluting the water in them. This is possible any time the pressure in the pipe is lower than that in any groundwater, or surface water, around the pipe.

There are a variety of pipe materials available, each with advantages and disadvantages. It is best to obtain local advice as to which material might be most suitable for the local conditions.

- **Metal pipes** are stronger than plastic ones and less prone to illegal connections, but they are prone to corrosion. They can resist high internal water pressures without bursting, and if necessary (e.g. in rocky areas), they can be laid on the surface.
- **PVC-u pipes** (often called uPVC) are fairly brittle, especially after they have been exposed to strong sunlight for a few months.
- **Polyethylene** (PE) pipes [high density (HDPE) and medium density (MDPE) pipes] are usually a good choice where available. Small diameter PE pipes can come in long rolls (e.g. 100 m) which reduces the number of joints needed and hence the potential for leakage. PE pipes can be joined by careful hand welding, using a simple heating plate, and this avoids the cost of pipe couplings which are often of the compression type. Also, since the pipe is quite flexible, it is easy to lay normal pipe to form bends.
- **Asbestos cement** pipes are used in some countries. There are no health risks from using this type of pipe for transmission of water. However, to avoid releasing asbestos dust when cutting and machining such pipes on site, some basic precautions are needed, such as keeping the pipe wet and burying all the cuttings.

7.4.2 Pipeline design

The size of pipe required depends on:

- the amount of water it has to carry
- the internal roughness of the pipe (which depends on the material it is made of, and, if it is prone to corrosion, the age of the pipe)
- the pressure at the inlet end of the pipe
- the difference in elevation between the inlet and outlet of the pipe.

When no water is flowing in a water-filled pipe distribution system the pressure at any point in the system is called the static pressure at that point. Water pressures are often measured in metres of water (called pressure head), relating to the height in metres to which water would rise in an imaginary vertical pipe connected to the point in question. Hence in a gravity flow system the maximum static head will be at the point where the height difference between the highest water level in the reservoir and the elevation of the pipework is a maximum (i.e. at the lowest point on the distribution system). The pressure here will be highest when there is no water flowing in the system (e.g. at night). In mountainous areas, gravity supply systems may need to be provided with intermediate 'break pressure tanks' along the route. At each tank, any pressure in the pipe is lost so this reduces the maximum static pressure downstream and the risk of burst pipes. Alternatively, different types of pipe may be used where the pressure is high.

When water is flowing along a pipe the pressure is less than that experienced during static conditions. This is because of friction between the water and the inside of the pipe. The pressure loss along a pipe increases with:

- increased flow rate
- reduced diameter
- increased length
- increased pipe roughness.

The challenge for the designer is to ensure that there is sufficient pressure at every tap to deliver water at a rate convenient to users. The system is usually designed to conveniently supply water during the time of peak demand (see Chapter 2). Usually the aim is to achieve at least 7 m of head at the tap when it is delivering water, although a lower head may be acceptable for short periods.

There are two basic types of distribution system (Figure 63), a branched system (which is laid out rather like the trunk and branches of a tree), and the looped or grid system (which is based on a grid of interconnected pipes). The latter is harder to design manually but has a number of advantages, particularly in more densely populated areas. These advantages generally include more stable water pressure for users, despite the use of smaller diameters of pipe, and the ease with which small sections of the system can be isolated for repairs without affecting the rest of it.

Design tables, charts and rules of thumb, and computer software are available to assist with choices of pipe diameter for distribution system design, but this is a subject which is too complicated to cover in this book. The flow rates chosen for each pipe should allow for future increases in population and increases in per capita demand during the life of the distribution system. Some of the sources in the appendices and references offer useful guidance on the design of piped systems.

7.4.3 Pipe laying

Different pipe materials require different methods of laying and jointing. With many types of pipe special precautions, such as concrete thrust blocks,

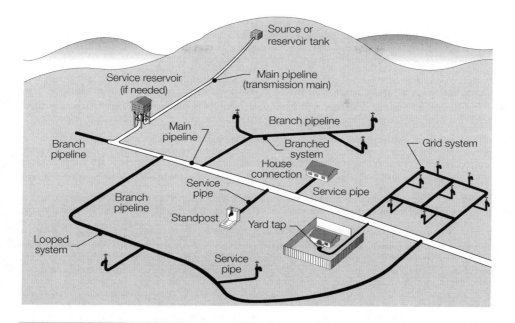

Figure 63 Pipework which makes up a distribution system.
Source: WEDC

may be necessary to restrain joints at bends. This is especially true with pipes that have push-fit joints. The advice of the pipe manufacturer should be followed, but a few general points can be made.

Pipes should be laid at least 0.3 m deep to the top of the pipe, 0.6 m generally and preferably 1 m deep when under roads or where there is a chance of their being exposed by ploughing or soil erosion. All the soil in contact with plastic pipes should be free of sharp stones, and be well-compacted in 100 mm layers by workers stamping on it with their feet or, preferably, hitting it with a flat, heavy object. This compaction is necessary so that the soil properly supports the flexible pipe. The soil placed around a metal pipe does not normally have to be free of stones, but it should still be well compacted. Ideally soil filling the whole trench above the pipe should be compacted in layers, particularly on slopes. This is to minimize settlement and to prevent the back-filled soil being washed away by surface water. The soil should be filled back to above the surface level, being built up to make a small mound to compensate for any settlement of the backfill which will later take place. If, because of poor compaction, or omission of the mound, a depression develops in the ground above the pipeline, surface water is likely to run along it and wash away the soil. If cuttings from thorn bushes are placed along the back-filled trench, this will help to keep goats away and allow grass to grow, protecting the soil from erosion. Another way to protect the trench from erosion is to put small logs or ridges of earth across it at intervals.

These make the surface water run off to one side instead of continuing along the trench eroding the backfill.

If pipes cannot be buried below ground, metal pipes should be used, but these may still need protection if they are in a place where animals or pedestrians may step on them. One form of protection is to surround the pipe with carefully placed stones. If only plastic pipes are available they should be laid in a soil embankment with gently sloping sides. In some situations, such as high altitude, pipes should be protected from frost.

Air valves may be needed at high points, to evacuate or admit air. If it is likely to be well used, the service pipe to a standpost can be connected to the top of the main pipe at a high point to evacuate any air. This will save the cost of, and maintenance problems associated with, an air valve.

Washout valves may be needed at low points to drain sections of pipe to carry out repairs or to flush out any solids deposited in the pipe. Valves should also be provided at appropriate points to allow sections of the pipe to be isolated for repair or extension work to be carried out.

At least once a year every pipeline should be checked for leaks and soil erosion. This will be easier if the route of the pipe has been marked with permanent markers. If the pipe is laid close to footpaths or tracks, leaks will be seen more quickly than if the pipe follows a more isolated route.

Appendix 1

Cross references to other books

This appendix is designed to help the user judge which of the six core reference books (listed at the start of the references and bibliography) are likely to be helpful as a source of further information about particular topics covered in this book. Other books are shown if the core reference books do not cover an important topic very well. Where the relevant information provided is of limited help the page numbers are shown in italics. Readers should note that Technical Briefs 33–64 are available on the internet at http://www.lboro.ac.uk/well/resources/technical-briefs/technical-briefs.htm.

Topics	Technical Brief No. (Appendix 2)	Relevant pages in the six core reference books (see start of references and bibliography)				Other useful references
		Smet & van Wijk (2002)	Jordan (1984)	Cairncross and Feachem (1983)	Watt & Wood (1979)	
IMPORTANT TOPICS NOT FULLY COVERED IN THIS BOOK						
Planning and management, and associated social, financial, human resource and institutional factors		16–45, 80–99	2–7, 15	47–61	39–49	WELL (1998), Dayal, van Wijk & Mukherjee (2000)
Choosing appropriate technology	55	21–23, 29–35		50		
TOPICS COVERED IN THIS BOOK (only sections that have corresponding sections in the core references are listed)						
1 Introduction						
1.2 Incremental approach		472–473				
1.3 Involving the community	30	16–45		56–58		IRC (1991), Dayal, van Wijk & Mukherjee (2000)
1.4 Importance of sanitation and hygiene	17, 19, 25, 32, 50, 51	16–19, 29		1–27, 52–53, 59–60	28–38	Almedom (1997), Boot (1991)
2 Design capacity		62–67, 473–481	27–29	52–53		
3 Sources of water	7, 49					
3.1.1 The main sources	55		7–8			
3.1.2 Judging the quality of water	18, 20, 50, 52	67–77, 145–146, 157–159, 560–565	13	29–35, 41	28–38	Lloyd and Helmer (1991)
3.2 Rainwater	11	130–149				Gould and Nissen-Petersen (1999)
3.3 Groundwater		102–127, 152–167, 200–250, 568–579				Price (1996)

Topics	Technical Brief No. (Appendix 2)	Relevant pages in the six core reference books				Other useful references
		Smet & van Wijk (2002)	Jordan (1984)	Cairncross and Feachem (1983)	Watt & Wood (1979)	
4.2.2 Human powered pumps	13, 33,35, 41	175–183				IRC (1988)
Reciprocating piston handpumps						
Other types of handpump and footpump		184–186, 190				
4.3 Wind powered pumps		171–172		77	212–214	Fraenkel et al (1993)
4.4 Water powered pumps		194–196	154–157	77		Fraenkel (1997)
4.5 Engine and motor powered pump		175–178, 186–194, 454–460		77–79		Fraenkel (1997), Barlow et al (1993)
4.5.1 Types of motorized pump		176–177, 186–194				
4.5.2 Suction limit and priming		179, 255–256				
4.5.3 Power sources for motorized pumps		170–174, 190–194				
4.5.4 Sustainability of motorized systems		170–174				
5 Storage	11, 14	473–477, 136–144		79–81		
5.2 Dams	48	121–125, 259–260				Nelson (1985)
5.3 Storage tanks		471–481, 132–144	124–139			Gould and Nissen-Petersen (1999)
5.3.3 Ferrocement tanks	36, 56	141				
5.3.4 Cement-mortar jars	1					
5.4 Storage in the home	1		176–177			

Section						Reference
6 Water treatment	58, 59	266–439, 500–531, 546–549	152–153	81–85, 89–106		Schulz and Okun (1984)
6.2 Screening	22, 47, 58	261–262, 542–543	102, 179–180			
6.3 Storage and sedimentation	58, 60	296–312, 314–326, 342, 544–546	114–119, 124–139	81–82, 90–92		IRC (1998) (for information on roughing filters)
6.4 Filtration				81, 91–94		
6.4.2 Rapid sand filters		360–389, 546–548				IRC (1998)
6.4.3 Slow sand filters	15, 21	328–358, 546–548	152			IRC (1998)
6.5 Disinfection	46, 58	412–439		81–84, 94–102		
6.5.2 Chemical disinfection		417–439, 548, 554			36–38, 99–100	
6.5.3 Solar disinfection		413–414				
6.5.4 Boiling		413				
6.6 Aeration and removal of iron, manganese, tastes and odours	58, 59	286–293	153	84–85	34–36	
6.7 Removal of salt	40	392–410				
7 Piped water distribution		442–497	1–224	86–87, 100–106		
7.2 House and yard connections		469		86		
7.3 Public water points	26	469–472, 549	140–146	86–87		
7.4 Pipeline materials, design and laying	29	458–497	1–224			

Appendix 2

Technical Briefs relating to water supply and associated issues

These Technical Briefs were originally published individually in *Waterlines* but are now published either in Pickford (1991) or Shaw (1999). Briefs from number 33 onwards can be downloaded from the WELL website (see Appendix 3). Each brief is a four-page summary of a particular topic with references to further sources of information.

No.	Title	Associated Technical Briefs (Numbers in italics are less relevant)
In Pickford (1991)		
1	Household water storage	11, 14, 36, 56
3	Protecting a spring	34
4	Lining a hand-dug well	39
5	Slotted bamboo tubewell screen	*43*
7	The water cycle	
11	Rainwater harvesting	1, 14, 36, 56
13	Handpumps	33, 35, 41
14	Above-ground rainwater storage	11, 36, 56
15	Slow sand filter design (1)	21, *1, 11, 15, 21, 47*
17	Health, water and sanitation (1)	19, *25, 50, 51*
18	Water testing	20
19	Health, water and sanitation (2)	17, *25, 50, 51*
20	Water sampling	18
21	Slow sand filters (2)	15, *1, 11, 15, 21, 47*
22	Intakes from rivers	*47*
24	Groundwater dams	
25	Eye and skin diseases	*17, 19*
26	Public standposts	*32*
27	Discharge measurements and estimates	*3, 34*
29	Designing simple pipelines	*44*
30	Community management	
32	Drainage for improved health	*26*
In Shaw (1999)		
33	Maintaining handpumps	13, 35, 41
34	Protecting springs—an alternative to spring boxes	3
35	Low-lift irrigation pumps	13, 33, 41
36	Ferrocement water tanks	11, 14, 56
39	Upgrading traditional wells	4
40	Desalination	*58, 59*

No.	Title	Associated Technical Briefs (Numbers in italics are less relevant)
41	VLOM pumps	13, 33, 35
43	Simple drilling methods	*5*
44	Emergency water supply	62
46	Chlorination	*58, 59*
47	Improving pond water	*22, 58, 59*
48	Small earth dams	*47*
49	Choosing an appropriate technology	55
50	Sanitary surveying	*17, 19, 51*
51	Water, sanitation and hygiene understanding	*17, 19, 50*
52	Water – quality or quantity	*17, 19*
55	Water source selection	49
56	Buried and semi-submerged water tanks	*11, 14, 36*
58	Household water treatment 1	59, *11, 15, 21, 40, 47, 60, 65*
59	Household water treatment 2	58, *11, 15, 21, 40, 47, 60, 65*
60	Water clarification using *Moringa oleifera* seed coagulant	*15, 21, 58, 59*
62	Emergency supply in cold regions	44

Appendix 3

Sources of information on the worldwide web

There are a number of websites relating to water supply in developing countries. Here are just a few of them, but most of these have useful links to many other relevant websites. The sites are listed below in three sections, General, Rainwater Catchment and Handpumps. A brief description of the relevant information available on each site at the time of preparing this book is shown before the site address.

General

GARNET

GARNET (Global Applied Research Network) is a mechanism for information exchange in the water supply and sanitation sector using low-cost, informal networks of researchers, practitioners and funders of research. It is a useful site giving links to a number of water and sanitation organizations, and other sites, networks and discussion groups related to specific topics, such as handpumps.

www.lboro.ac.uk/garnet

Lifewater Canada

This very useful site gives information about and links to relevant information about many aspects relating to low-cost water supply. You will find many relevant links by following the 'Online Training and Technical Links' option. The 'Water for the World' technical notes are a useful resource.

www.lifewater.ca

SKAT

The SKAT (Swiss Centre for Development Cooperation in Technology and Management) website also has useful links to a number of sources of further information relating to water supply. The 'technologies' section is found at www.skat.ch/watsanweb/content/technology/general.htm, but other sections are also useful.

www.skat.ch/watsanweb

Waterlines

This is the site of *Waterlines*, the International Journal of Appropriate Technologies for Water Supply and Sanitation. The index to past magazines can help you to locate relevant articles if you can get access to back copies of the magazine, or order them from ITDG Publishing.

www.itdgpublishing.org.uk/waterlines.htm

WELL

WELL is a resource centre promoting environmental health and well-being in developing and transitional countries. The Centre is funded by the United Kingdom's Department for International Development (DFID). WELL provides a free Immediate Technical Response service to enquiries from NGOs, UK government and UN organizations. The site includes copies of all Waterline Technical Briefs from No.33 (see Appendix 2) which are from *Running Water*, one of the six core reference books. These Briefs are cross-referred to in Appendix 1. WELL also provides a searchable library catalogue that is useful for finding details about specialist books relating to water, environmental sanitation, health and hygiene in developing countries. Links are provided to many other sites.

www.lboro.ac.uk/well

World Bank

The following page on the World Bank website links to useful sources of information on water supply technologies.

www.worldbank.org/html/fpd/water/topics/tech_supply.html

Rainwater catchment

DTU Domestic Roofwater Harvesting Programme

A useful site from the Development Technology Unit of the School of Engineering, University of Warwick, with lots of links to other sites related to the catchment and storage of rainwater.

www.eng.warwick.ac.uk/DTU/rwh/index.html

Handpumps

HTN

HTN, the global Network for Cost-effective Technologies in Water Supply, coordinates and facilitates resolution of outstanding design issues and identifies priority areas for research and development of handpumps. It maintains and disseminates the international specifications of public domain handpumps, such as the Afridev, Uganda U3, Bush, MALDA, and the Yaku-MAYA-TARA pumps and others. It also promotes information sharing and capacity building.

www.skat.ch/htn

Lifewater Canada

This site, mentioned above, is a very useful site giving information about and links to other sites relating to many types of handpumps and footpumps. You will find these by clicking on the 'Online Training Manuals and Technical Links' option.

www.lifewater.ca

Appendix 4

Units of measurement

Here are a few conversion factors that you may find useful:

Length: 1 metre, 1 m = 100 cm = 1000 mm = 3.281 feet = 1.094 yards
1 kilometre, 1 km = 1000 m = 0.6214 miles
1 inch, 1″ = 25.4 mm = 2.54 cm
1 foot, 1 ft = 304.8 mm = 30.48 cm
1 mile = 1.609 km

Mass: 1 kg = 1000 g = 1 000 000 mg
1 kg = 2.204 lb = 35.27 oz

Volume: $1 \ m^3$ = 1000 litres
$1 \ m^3$ = 35.31 ft^3 = 220.0 British gallons = 264.2 US gallons
1 litre = 0.22 British gallons = 0.264 US gallons
1 megalitre, 1 Ml = 1000 m^3
1 British gallon = 1.2 US gallons = 0.1605 ft^3 = 4.546 litres

Pressure: 1 metre head of water = 0.1 kgf/cm^2 = 0.00981 N/mm^2
1 metre head of water = 1.422 lb/in^2 = 1.422 psi
1 Pascal, 1 Pa = 1 N/m^2
1 bar = 10.197 metres head of water (although often for simplicity it is assumed to be 10 m).

Flow rates: 1 litre/s = 13.20 British gallons per minute (gpm)
1 m^3/h = 0.001 Ml/h
1 m^3/h = 0.00527928 million British gallons per day (mgd)

References and bibliography

Six core references

Cairncross, S. and Feachem, R. (1993) *Environmental Health Engineering in the Tropics: An introductory text*, second edition, Wiley, Chichester, UK, 310pp, ISBN 0 471 93885 8
A number of sections in this book give a good introduction to the diseases related to water supply and sanitation and possible methods of control. It contains useful sections on rural water supply and sanitation. The section on water supply duplicates some of the material in the original Ross Bulletin No.10 published by the same authors in 1986.

Jordan, T.D. Jr. (1984) *A Handbook of Gravity-Flow Water Systems for Small Communities*, ITDG Publishing, London, UK, 250pp, ISBN 0 94668 850 8
This is a most comprehensive book on the subject and is recommended to anyone who plans to implement piped gravity-flow systems.

Pickford, J. (ed.) (1991) *The Worth of Water: Technical briefs on health, water and sanitation*, ITDG Publishing, London, UK. ISBN 1 85339 069 0
This book contains Technical Briefs 1–32 that were originally published singly in *Waterlines*. The Briefs are useful, well illustrated, four-page summaries of the important aspects relating to various aspects of health, water supply and sanitation in developing countries. They contain references to important sources of further information. Note that Technical Briefs 33–64 are published in Shaw (ed.) (1999).

Shaw, R.J. (ed.) (1999) *Running Water: More technical briefs on health, water and sanitation*, ITDG Publishing, London, UK, ISBN 1 85339 450 5
This book contains Technical Briefs 33–64 that were originally published singly in *Waterlines*. The Briefs are useful, well illustrated, four-page summaries of the important aspects relating to various aspects of health, water supply and sanitation in developing countries. Note that Technical Briefs 1–32 are published in Pickford (ed.) (1991). Technical Briefs 33–64 are available at www.lboro.ac.uk/well/resources/technical-briefs/technical-briefs.htm

Smet J. and van Wijk C. (eds) (2002) *Small Community Water Supplies: Technology, people and partnerships*, IRC Technical Paper Series 40, IRC International Water and Sanitation Centre, Delft, The Netherlands. ISBN 90 6687 035 4
This is an updated version of *Small Community Water Supplies: Technology of small water supply systems in developing countries* published in 1981. It has fairly comprehensive coverage of the technological aspects of water supplies. This latest version has a new chapters relating to integrated water resources management and water supply in disasters and emergencies. Other new chapters cover desalination, fluoride removal and arsenic removal. There is a revised annex on sanitary surveys and laboratory analysis, and a new annex on groundwater exploration. Chapters 1–4 are available at www.irc.nl/products/publications/online/tp40e/index.html

Watt, S.B. and Wood, W.E. (1979) *Hand Dug Wells and their Construction* (second edition), ITDG Publishing, London, UK, 254pp, ISBN 0 90303 127 2
This is a most comprehensive book on the subject and is recommended to any who plan to implement hand dug wells.

Other relevant publications

The following titles deal with subjects not covered, or covered only briefly, in the six core books.

Almedom, A.M., Blumenthal, U. and Manderson, L. (1997) *Hygiene Evaluation Procedures: Approaches and methods for assessing water- and sanitation-related hygiene practices*, International Nutritional Foundation for Developing Countries, PO Box 500, Boston MA, 02114-0500, USA, ISBN 0 9635522 8 7
This book provides practical guidelines for evaluating water- and sanitation-related hygiene practices. It can be read at: www.unu.edu/unupress/food2/uin11e/uin11e00.htm

Barlow, R., McNelis, B. and Derrick, A. (1993) *Solar Pumping: An introduction and update on the technology, performance, cost and economics*, ITDG Publishing, London, UK ISBN 1 85339 179 4, and World Bank, Washington, DC, USA, ISBN 0 8213 2101 3
A comprehensive book which includes a comparison of the cost of solar, diesel, hand and wind power for pumping water.

Boot, M.T. (1991) *Just Stir Gently: The way to mix hygiene education with water supply and sanitation*, Technical Paper Series No.29, International Water and Sanitation Centre (IRC), The Hague, The Netherlands, ISBN 90 6687 016 8
A useful book which looks at all aspects of understanding and influencing health behaviour relating to water and sanitation.

Dayal R., van Wijk C. & Mukherjee N. (2000) *Methodology for Participatory Assessments with Communities, Institutions and Policy Makers: Linking sustainability with demand, gender and poverty*, Metguide, Water and Sanitation Program, The World Bank, Washington DC, USA and IRC International Water and Sanitation Centre, Delft, The Netherlands.
An excellent book that describes participatory field 'tools' that can be used to help ensure that water and sanitation projects meet users' needs in the most appropriate way. It is available at www.wsp.org/pdfs/global_metguideall.pdf

Fraenkel, P., Barlow, R., Crick, F., Derrick, A. and Bokalders, V. (1993) *Windpumps: A guide for development workers*, ITDG Publishing, London, UK, in association with the Stockholm Environmental Institute, 156pp, ISBN 1 85339 126 3
A comprehensive book which gives details about a number of windpump manufacturers and suppliers.

Fraenkel, P. (1997) *Water-pumping Devices: A handbook for users and choosers*, (second edition), ITDG Publishing, London, UK, 254pp, ISBN 1 85339 346 0
This book covers all types of pumping systems driven by manual, animal, wind, water, solar, engine and motor power.

Gould, J. and Nissen-Petersen, E. (1999) *Rainwater Catchment Systems for Domestic Supply: Design, construction and implementation*, ITDG Publishing, London, UK. ISBN 1 85339 456 4
A very comprehensive and up-to-date book on the subject of rainwater catchment and storage. It includes a good section that gives outline details for the construction of a number of different types of storage tank.

IRC (1988) *Handpumps: Issues and concepts in rural water supply programmes*, IRC Technical Paper Series No.25, International Water and Sanitation Centre, The Hague, The Netherlands, 163pp, ISBN 90 6687 010 9
A book that gives useful advice on handpump technology, installation, maintenance and manufacture.

IRC (1998) *Multi-Stage Filtration: An innovative water treatment technology*, Technical Paper Series 34-E, International Water and Sanitation Centre, The Hague, The Netherlands and CINARA, Valle University, Calí, Colombia,165pp, ISBN 90 6687 028 1
A book which gives guidance on how to treat surface water using different types of roughing filter together with slow sand filters.

IRC (1991) *Partners for Progress: An approach to sustainable piped water supplies*, Technical Paper Series 28, International Water and Sanitation Centre, The Hague, The Netherlands, 139pp, ISBN 90 6687 015X

A good introduction to the philosophy and practice of full involvement of communities in the water supplies designed to serve them. Although the focus is on piped schemes, many of the suggestions are also relevant to other water supply methods. The book covers planning, implementation, operation, maintenance, monitoring and evaluation.

Lloyd, B. and Helmer, R. (1991) *Surveillance of Drinking Water Quality in Rural Areas*, Longman Scientific & Technical, UK, ISBN 0 582 06330 2

In addition to information about water quality testing this book describes 'sanitary surveying'. This technique can be used to judge the likelihood of water sources being contaminated without the need to carry out bacteriological testing.

Morgan, P. (1990) *Rural Water Supplies and Sanitation*, Macmillan, London, UK, 358pp, ISBN 0 33348 569 6

Two thirds of this book deals with practical aspects of exploiting groundwater and the other third is about low-cost sanitation. The water supply section has particular emphasis on use of the Blair bucket pump and various types of handpump made and used in Zimbabwe. Other sections provides some guidance on constructing hand-dug wells and hand-augered boreholes using the Vonder rig. It has short sections on the purification of water and on rainwater harvesting.

Nelson, K.D. (1985) *Design and Construction of Small Earth Dams*, Heinemann, UK, 128pp, ISBN 0 90960 534 3

An introduction to safe construction of earth dams.

Nilsson, A. (1988) *Groundwater Dams for Small Water Supply*, ITDG Publishing, London, UK, 64pp, ISBN 1 85339 050 X.

A good reference book on groundwater dams.

Price, M. (1996) *Introducing Groundwater* (second edition), Chapman & Hall, London, UK, 278pp ISBN 0 412 48500 1

A good introductory book on the occurrence and quality of groundwater, the way in which it flows, how it can be located and tested, and the construction and use of boreholes.

Schulz, C.R. and Okun, D.A. (1992) *Surface Water Treatment for Communities in Developing Countries*, ITDG Publishing. ISBN 1 85339 142 5

A comprehensive book dealing with most aspects of the design of small and large scale water treatment plants. This book is a reprint of one first published in 1984. In places it is now a little out of date but it is still very useful.

WELL (1998) *DFID Guidance Manual on Water Supply and Sanitation Programmes*, WELL Resource Centre Network for Water, Sanitation and Environmental Health, Water, Engineering and Development Centre (WEDC), Loughborough University, UK.

A comprehensive manual prepared by WELL, a resource centre funded by DFID. It covers the whole of the project cycle for water and sanitation programmes (i.e. from policy development to evaluation) dealing mainly with important programme planning and management issues rather than technology. It can be read online, printed or ordered, via the WELL website at www.lboro.ac.uk/well/resources/books-and-manuals/guidance-manual.htm

Other works cited

DHV (1985) *Low Cost Water Supply: For human consumption, cattle watering, small scale irrigation*, Part 1: Survey and Construction of Wells, DHV Consulting Engineers, Amersfoort, The Netherlands.

DLVW (1983) *Rural Water Operators Handbook: Gravity fed rural pipe water schemes*, Department of Lands, Valuation and Water, Government of Malawi, Malawi.

Guoth-Gumberger, M. and R. (1987) *Small Projects Training Manual*, Volume II, Water Supply, Sudan Council of Churches, Munuki Water and Sanitation Project, Sudan.

Hasse, R. (1989) *Rainwater Reservoirs, Above Ground Structures for Roof Catchment*, Friedr. Vieweg & Sohn Verlagsgesellschaft mbH, Braunschweig, Germany.

IRC (1979) *Public Standpost Water Supplies: A design manual*, IRC Technical Paper No.14, International Water and Sanitation Centre, The Hague, The Netherlands.

IRC (1987) *Slow Sand Filtration for Community Water Supply: Planning, design, construction, operation and maintenance*, Technical Paper No.24, International Water and Sanitation Centre, The Hague, The Netherlands.

IRC (1988) *Community Self-Improvement in Water Supply and Sanitation: A training and reference manual for community health workers, community development workers and other community based workers*, Training Series No.5, International Water and Sanitation Centre (IRC), The Hague, The Netherlands.

Nissen-Petersen, E. and Lee, M. (1990) *Harvesting Rainwater in Semi-Arid Africa*, Manual No.3, Rock Catchment Dam with Self Closing Tap, ASAL Rainwater Harvesting, Nairobi, Kenya.

Rajagopalan, S. and Shiffman, M.A. (1974) *Guide to Simple Sanitary Measures for the Control of Enteric Diseases*, World Health Organization, Geneva.

SWS (1992) Trade literature from SWS Filtration Ltd, Morpeth, Northumberland, UK.

Wegelin, M. (1996) *Surface Water Treatment by Roughing Filters: A design, construction and operation manual*, SANDEC Report No.2/96, Swiss Centre for Development Cooperation in Technology and Management (SKAT), St. Gallen, Switzerland.